CAMBRIDGE LIBRARY COLLECTION

Books of enduring scholarly value

Philosophy

This series contains both philosophical texts and critical essays about philosophy, concentrating especially on works originally published in the eighteenth and nineteenth centuries. It covers a broad range of topics including ethics, logic, metaphysics, aesthetics, utilitarianism, positivism, scientific method and political thought. It also includes biographies and accounts of the history of philosophy, as well as collections of papers by leading figures. In addition to this series, primary texts by ancient philosophers, and works with particular relevance to philosophy of science, politics or theology, may be found elsewhere in the Cambridge Library Collection.

Collected Essays

Known as 'Darwin's Bulldog', the biologist Thomas Henry Huxley (1825–95) was a tireless supporter of the evolutionary theories of his friend Charles Darwin. Huxley also made his own significant scientific contributions, and he was influential in the development of science education despite having had only two years of formal schooling. He established his scientific reputation through experiments on aquatic life carried out during a voyage to Australia while working as an assistant surgeon in the Royal Navy; ultimately he became President of the Royal Society (1883–5). Throughout his life Huxley struggled with issues of faith, and he coined the term 'agnostic' to describe his beliefs. This nine-volume collection of Huxley's essays, which he edited and published in 1893–4, demonstrates the wide range of his intellectual interests. In Volume 5, Huxley discusses the doctrines of Christianity and explains how his dissatisfaction with conventional religion led him to agnosticism.

Cambridge University Press has long been a pioneer in the reissuing of out-of-print titles from its own backlist, producing digital reprints of books that are still sought after by scholars and students but could not be reprinted economically using traditional technology. The Cambridge Library Collection extends this activity to a wider range of books which are still of importance to researchers and professionals, either for the source material they contain, or as landmarks in the history of their academic discipline.

Drawing from the world-renowned collections in the Cambridge University Library, and guided by the advice of experts in each subject area, Cambridge University Press is using state-of-the-art scanning machines in its own Printing House to capture the content of each book selected for inclusion. The files are processed to give a consistently clear, crisp image, and the books finished to the high quality standard for which the Press is recognised around the world. The latest print-on-demand technology ensures that the books will remain available indefinitely, and that orders for single or multiple copies can quickly be supplied.

The Cambridge Library Collection will bring back to life books of enduring scholarly value (including out-of-copyright works originally issued by other publishers) across a wide range of disciplines in the humanities and social sciences and in science and technology.

Collected Essays

VOLUME 5: SCIENCE AND THE CHRISTIAN TRADITION

THOMAS HENRY HUXLEY

CAMBRIDGE UNIVERSITY PRESS

Cambridge, New York, Melbourne, Madrid, Cape Town,
Singapore, São Paolo, Delhi, Tokyo, Mexico City

Published in the United States of America by Cambridge University Press, New York

www.cambridge.org
Information on this title: www.cambridge.org/9781108040556

© in this compilation Cambridge University Press 2011

This edition first published 1894
This digitally printed version 2011

ISBN 978-1-108-04055-6 Paperback

COLLECTED ESSAYS

By T. H. HUXLEY

VOLUME V

SCIENCE

AND

CHRISTIAN TRADITION

ESSAYS

BY

THOMAS H HUXLEY

London

MACMILLAN AND CO.

1894

PREFACE

"For close upon forty years I have been writing with one purpose; from time to time, I have fought for that which seemed to me the truth, perhaps still more, against that which I have thought error; and, in this way, I have reached, indeed over-stepped, the threshold of old age. There, every earnest man has to listen to the voice within: 'Give an account of thy stewardship, for thou mayest be no longer steward.'

"That I have been an unjust steward my conscience does not bear witness. At times blundering, at times negligent, Heaven knows: but, on the whole, I have done that which I felt able and called upon to do; and I have done it without looking to the right or to the left; seeking no man's favour, fearing no man's disfavour.

"But what is it that I have been doing? In the end one's conceptions should form a whole, though only parts may have found utterance, as occasion arose; now do these exhibit harmony and mutual connexion? In one's zeal much of the old gets broken to pieces; but has one made ready something new, fit to be set in the place of the old?

"That they merely destroy without reconstructing, is the especial charge, with which those who work in this direction are constantly reproached. In a certain sense I do not defend myself against the charge; but I deny that any reproach is deserved.

"I have never proposed to myself to begin outward construc-tion; because I do not believe that the time has come for it. Our present business is with inward preparation, especially the

preparation of those who have ceased to be content with the
old, and find no satisfaction in half measures. I have wished,
and I still wish, to disturb no man's peace of mind, no man's
beliefs; but only to point out to those in whom they are
already shattered, the direction in which, in my conviction,
firmer ground lies." [1]

So wrote one of the protagonists of the New
Reformation—and a well-abused man if ever
there was one—a score of years since, in the remarkable book in which he discusses the negative
and the positive results of the rigorous application
of scientific method to the investigation of the
higher problems of human life.

Recent experience leads me to imagine that
there may be a good many countrymen of my
own, even at this time, to whom it may be profit-
able to read, mark and inwardly digest, the
weighty words of the author of that " Leben Jesu,"
which, half a century ago, stirred the religious
world so seriously that it has never settled down
again quite on the old foundations; indeed, some
think it never will. I have a personal interest in
the carrying out of the recommendation I venture
to make. It may enable many worthy persons, in
whose estimation I should really be glad to stand
higher than I do, to become aware of the possibility
that my motives in writing the essays, contained
in this and the preceding volume, were not exactly
those that they ascribe to me.

[1] D. F. Strauss, *Der alte und der neue Glaube* (1872), pp. 9-10.

I too have reached the term at which the still, small voice, more audible than any other to the dulled ear of age, makes its demand; and I have found that it is of no sort of use to try to cook the accounts rendered. Nevertheless, I distinctly decline to admit some of the items charged; more particularly that of having "gone out of my way" to attack the Bible; and I as steadfastly deny that "hatred of Christianity" is a feeling with which I have any acquaintance. There are very few things which I find it permissible to hate; and though, it may be, that some of the organisations, which arrogate to themselves the Christian name, have richly earned a place in the category of hateful things, that ought to have nothing to do with one's estimation of the religion, which they have perverted and disfigured out of all likeness to the original.

The simple fact is that, as I have already more than once hinted, my story is that of the wolf and the lamb over again. I have never "gone out of my way" to attack the Bible, or anything else: it was the dominant ecclesiasticism of my early days, which, as I believe, without any warrant from the Bible itself, thrust the book in my way.

I had set out on a journey, with no other purpose than that of exploring a certain province of natural knowledge; I strayed no hair's breadth from the course which it was my right and my duty to pursue; and yet I found that, whatever route I took, before long, I came to a tall and formidable-looking fence. Confident as I might be in the existence of an ancient and indefeasible right of way, before me stood the thorny barrier with its comminatory notice-board—" No Thoroughfare. By order. Moses." There seemed no way over; nor did the prospect of creeping round, as I saw some do, attract me. True there was no longer any cause to fear the spring guns and man-traps set by former lords of the manor; but one is apt to get very dirty going on all-fours. The only alternatives were either to give up my journey—which I was not minded to do—or to break the fence down and go through it.

Now I was and am, by nature, a law-abiding person, ready and willing to submit to all legitimate authority. But I also had and have a rooted conviction, that reasonable assurance of the legitimacy should precede the submission ; so I made it my business to look up the manorial title-deeds. The pretensions of the ecclesiastical "Moses" to exercise a control over the operations of the reasoning faculty in the search after truth, thirty centuries after his age, might be justifiable ; but, assuredly, the credentials produced in justification of claims so large required careful scrutiny.

Singular discoveries rewarded my industry. The ecclesiastical "Moses" proved to be a mere traditional mask, behind which, no doubt, lay the features of the historical Moses—just as many a mediæval fresco has been hidden by the whitewash of Georgian churchwardens. And as the æsthetic rector too often scrapes away the defacement, only to find blurred, parti-coloured patches, in which the original design is no longer to be traced; so, when the successive layers of Jewish and Christian traditional pigment, laid on, at intervals, for near three thousand years, had been removed, by even the tenderest critical operations, there was not much to be discerned of the leader of the Exodus.

Only one point became perfectly clear to me, namely, that Moses is not responsible for nine-tenths of the Pentateuch ; certainly not for the legends which had been made the bugbears of science. In fact, the fence turned out to be a mere heap of dry sticks and brushwood, and one might walk through it with impunity : the which I did. But I was still young, when I thus ventured to assert my liberty ; and young people are apt to be filled with a kind of *sæva indignatio*, when they discover the wide discrepancies between things as they seem and things as they are. It hurts their vanity to feel that they have prepared themselves for a mighty struggle to climb over, or break their way through, a rampart, which turns out, on close approach, to be a mere heap of ruins ; venerable, indeed, and archæologically interesting, but of no other moment. And some fragment of the super-fluous energy accumulated is apt to find vent in strong language.

Such, I suppose, was my case, when I wrote some passages which occur in an essay reprinted among "Darwiniana."[1] But when, not long ago "the voice" put it to me, whether I had better not expunge, or modify, these passages; whether, really, they were not a little too strong; I had to reply, with all deference, that while, from a merely literary point of view, I might admit them to be rather crude, I must stand by the substance of these items of my expenditure. I further ventured to express the conviction that scientific criticism of the Old Testament, since 1860, has justified every word of the estimate of the authority of the ecclesiastical "Moses" written at that time. And, carried away by the heat of self-justification, I even ventured to add, that the desperate attempt now set afoot to force biblical and post-biblical mythology into elementary instruction, renders it useful and necessary to go on making a considerable outlay in the same direction. Not yet, has "the cosmogony of the semi-barbarous Hebrew" ceased to be the "incubus of the philosopher, and the opprobrium of the orthodox;" not yet, has "the zeal of the Bibliolater" ceased from troubling; not yet, are the weaker sort, even of the instructed, at rest from their fruitless toil "to harmonise impossibilities," and "to force the generous new wine of science into the old bottles of Judaism."

But I am aware that the head and front of my

[1] *Collected Essays*, vol. ii., "On the Origin of Species" (1860).

offending lies not now where it formerly lay. Thirty
years ago, criticism of " Moses " was held by most
respectable people to be deadly sin; now it has
sunk to the rank of a mere peccadillo ; at least, if
it stops short of the history of Abraham. Destroy
the foundation of most forms of dogmatic Christianity contained in the second chapter of Genesis, if
you will; the new ecclesiasticism undertakes to
underpin the superstructure and make it, at any rate
to the eye, as firm as ever: but let him be anathema
who applies exactly the same canons of criticism
to the opening chapters of " Matthew " or of
" Luke." School-children may be told that the
world was by no means made in six days, and that
implicit belief in the story of Noah's Ark is permissible only, as a matter of business, to their
toy-makers ; but they are to hold for the certainest
of truths, to be doubted only at peril of their
salvation, that their Galilean fellow-child Jesus,
nineteen centuries ago, had no human father

Well, we will pass the item of 1860, said "the
voice." But why all this more recent coil about
the Gadarene swine and the like? Do you pretend that these poor animals got in your way,
years and years after the " Mosaic " fences were
down, at any rate so far as you are concerned?

Got in my way? Why, my good " voice," they
were driven in my way. I had happened to
make a statement, than which, so far as I have

ever been able to see, nothing can be more modest or inoffensive; to wit, that I am convinced of my own utter ignorance about a great number of things, respecting which the great majority of my neighbours (not only those of adult years, but children repeating their catechisms) affirm themselves to possess full information. I ask any candid and impartial judge, Is that attacking anybody or anything?

Yet, if I had made the most wanton and arrogant onslaught on the honest convictions of other people, I could not have been more hardly dealt with. The pentecostal charism, I believe, exhausted itself amongst the earliest disciples. Yet any one who has had to attend, as I have done, to copious objurgations, strewn with such appellations as "infidel" and "coward," must be a hardened sceptic indeed if he doubts the existence of a "gift of tongues" in the Churches of our time; unless, indeed, it should occur to him that some of these outpourings may have taken place after "the third hour of the day." I am far from thinking that it is worth while to give much attention to these inevitable incidents of all controversies, in which one party has acquired the mental peculiarities which are generated by the habit of much talking, with immunity from criticism. But as a rule, they are the sauce of dishes of misrepresentations and inaccuracies which it may be a duty, nay, even an innocent pleasure, to expose. In the particular case of which I am thinking, I felt, as Strauss says, "able and called upon" to undertake the business : and it is no responsibility of mine, if I found the Gospels, with their miraculous stories, of which the Gadarene is a typical example, blocking my way, as heretofore, the Pentateuch had done.

I was challenged to question the authority for the theory of "the spiritual world," and the practical consequences deducible from human relations to it, contained in these documents.

In my judgment, the actuality of this spiritual world—the value of the evidence for its objective existence and its influence upon the course of things—are matters, which lie as much within the province of science, as any other question about the existence and powers of the varied forms of living and conscious activity.

It really is my strong conviction that a man has no more right to say he believes this world is haunted by swarms of evil spirits, without being able to produce satisfactory evidence of the fact, than he has a right to say, without adducing adequate proof, that the circumpolar antarctic ice swarms with sea-serpents. I should not like to assert positively that it does not. I imagine that no cautious biologist would say as much ; but while quite open to conviction, he might properly decline to waste time upon the consideration of talk, no better accredited than forecastle

" yarns," about such monsters of the deep.
And if the interests of ordinary veracity dictate
this course, in relation to a matter of so little
consequence as this, what must be our obligations
in respect of the treatment of a question which is
fundamental alike for science and for ethics ? For
not only does our general theory of the universe
and of the nature of the order which pervades it,
hang upon the answer ; but the rules of practical
life must be deeply affected by it.

The belief in a demonic world is inculcated
throughout the Gospels and the rest of the books
of the New Testament ; it pervades the whole
patristic literature ; it colours the theory and the
practice of every Christian church down to modern
times.   Indeed, I doubt, if even now, there is
any church which, officially, departs from such a
fundamental doctrine of primitive Christianity as
the existence, in addition to the Cosmos with
which natural knowledge is conversant, of a world
of spirits ; that is to say, of intelligent agents, not
subject to the physical or mental limitations of
humanity, but nevertheless competent to interfere,
to an undefined extent, with the ordinary course of
both physical and mental phenomena.

More especially is this conception fundamental
for the authors of the Gospels. Without the belief
that the present world, and particularly that part
of it which is constituted by human society, has
been given over, since the Fall, to the influence

of wicked and malignant spiritual beings, governed and directed by a supreme devil—the moral antithesis and enemy of the supreme God— their theory of salvation by the Messiah falls to pieces. "To this end was the Son of God manifested, that he might destroy the works of the devil." [1]

The half-hearted religiosity of latter-day Christianity may choose to ignore the fact; but it remains none the less true, that he who refuses to accept the demonology of the Gospels rejects the revelation of a spiritual world, made in them, as much as if he denied the existence of such a person as Jesus of Nazareth; and deserves, as much as any one can do, to be ear-marked "infidel" by our gentle shepherds.

Now that which I thought it desirable to make perfectly clear, on my own account, and for the sake of those who find their capacity of belief in the Gospel theory of the universe failing them, is the fact, that, in my judgment, the demonology of primitive Christianity is totally devoid of foundation; and that no man, who is guided by the rules of investigation which are found to lead to the discovery of truth in other matters, not merely of science, but in the everyday affairs of life, will arrive at any other conclusion. To those who profess to be otherwise guided, I have nothing

[1] 1 John iii. 8.

to say; but to beg them to go their own way and leave me to mine.

I think it may be as well to repeat what I have said, over and over again, elsewhere, that *a priori* notions, about the possibility, or the impossibility, of the existence of a world of spirits, such as that presupposed by genuine Christianity, have no influence on my mind. The question for me is purely one of evidence : is the evidence adequate to bear out the theory, or is it not ? In my judgment it is not only inadequate, but quite absurdly insufficient. And on that ground, I should feel compelled to reject the theory; even if there were no positive grounds for adopting a totally different conception of the Cosmos.

For most people, the question of the evidence of the existence of a demonic world, in the long run, resolves itself into that of the trustworthiness of the Gospels; first, as to the objective truth of that which they narrate on this topic; second, as to the accuracy of the interpretation which their authors put upon these objective facts. For example, with respect to the Gadarene miracle, it is one question whether, at a certain time and place, a raving madman became sane, and a herd of swine rushed into the lake of Tiberias; and quite another, whether the cause of these occurrences was the transmigration of certain devils from the man into the pigs. And again, it is one question whether Jesus made a long oration on a certain occasion, mentioned in the first Gospel; altogether another, whether more or fewer of the propositions contained in the "Sermon on the Mount" were uttered on that occasion. One may give an affirmative answer to one of each of these pairs of questions and a negative to the other: one may affirm all, or deny all.

In considering the historical value of any four documents, proof when they were written and who wrote them is, no doubt, highly important. For if proof exists, that A B C and D wrote them, and that they were intelligent persons, writing independently and without prejudice, about facts within their own knowledge—their statements must need be worthy of the most attentive consideration.[1] But, even ecclesiastical tradition does not assert that either "Mark" or "Luke" wrote from his own knowledge—indeed "Luke" expressly asserts he did not. I cannot discover that any competent authority now maintains that the apostle Matthew wrote the Gospel which passes under his name. And whether the apostle John had, or had not, anything to do with the fourth Gospel; and if he had, what his share amounted to; are, as everybody who has attended to these matters knows, questions still hotly disputed, and with regard to which the extant evidence can

---

[1] Not necessarily of more than this. A few centuries ago the twelve most intelligent and impartial men to be found in England, would have independently testified that the sun moves, from east to west, across the heavens every day.

hardly carry no impartial judge beyond the admission of a possibility this way or that.

Thus, nothing but a balancing of very dubious probabilities is to be attained by approaching the question from this side. It is otherwise if we make the documents tell their own story : if we study them, as we study fossils, to discover internal evidence of when they arose, and how they have come to be. That really fruitful line of inquiry has led to the statement and the discussion of what is known as the *Synoptic Problem.*

In the Essays (VII.—XI.) which deal with the consequences of the application of the agnostic principle to Christian Evidences, contained in this volume, there are several references to the results of the attempts which have been made, during the last hundred years, to solve this problem. And, though it has been clearly stated and discussed, in works accessible to, and intelligible by, every English reader,[1] it may be well that I should here set forth a very brief exposition of the matters of fact out of which the problem has arisen ; and of some consequences, which, as I conceive, must be admitted if the facts are accepted.

These undisputed and, apparently, indisputable data may be thus stated :

I. The three books of which an ancient, but

---

[1] Nowhere more concisely and clearly than in Dr. Sutherland Black's article "Gospels" in Chambers's *Encyclopædia.* References are given to the more elaborate discussions of the problem

very questionable, ecclesiastical tradition asserts
Matthew, Mark, and Luke to be the authors,
agree, not only in presenting the same general
view, or *Synopsis*, of the nature and the order
of the events narrated; but, to a remarkable
extent, the very words which they employ
coincide.

II. Nevertheless, there are many equally marked,
and some irreconcilable, differences between
them. Narratives, verbally identical in some portions, diverge more or less in others. The order
in which they occur in one, or in two, Gospels
may be changed in another. In "Matthew" and
in "Luke" events of great importance make their
appearance, where the story of "Mark" seems to
leave no place for them; and, at the beginning and
the end of the two former Gospels, there is a
great amount of matter of which there is no
trace in "Mark."

III. Obvious and highly important differences,
in style and substance, separate the three
"Synoptics," taken together, from the fourth
Gospel, connected, by ecclesiastical tradition, with
the name of the apostle John. In its philosophical
proemium; in the conspicuous absence of exorcistic
miracles; in the self-assertive theosophy of the
long and diffuse monologues, which are so utterly
unlike the brief and pregnant utterances of Jesus
recorded in the Synoptics; in the assertion that
the crucifixion took place before the Passover,

which involves the denial, by implication, of the truth of the Synoptic story—to mention only a few particulars—the " Johannine " Gospel presents a wide divergence from the other three.

IV. If the mutual resemblances and differences of the Synoptic Gospels are closely considered, a curious result comes out; namely, that each may be analyzed into four components. The *first* of these consists of passages, to a greater or less extent verbally identical, which occur in all three Gospels. If this triple tradition is separated from the rest it will be found to comprise:

a. A narrative, of a somewhat broken and anecdotic aspect, which covers the period from the appearance of John the Baptist to the discovery of the emptiness of the tomb, on the first day of the week, some six-and-thirty hours after the crucifixion.

b. An apocalyptic address.

c. Parables and brief discourses, or rather, centos of religious and ethical exhortations and injunctions.

The *second* and the *third* set of components of each Gospel present equally close resemblances to passages, which are found in only one of the other Gospels; therefore it may be said that, for them, the tradition is double. The *fourth* component is peculiar to each Gospel; it is a single tradition and has no representative in the others.

To put the facts in another way: each Gospel

is composed of a *threefold tradition*, two *twofold traditions*, and one *peculiar tradition*. If the Gospels were the work of totally independent writers, it would follow that there are three witnesses for the statements in the first tradition; two for each of those in the second, and only one for those in the third.

V. If the reader will now take up that extremely instructive little book, Abbott and Rushbrooke's "Common Tradition" he will easily satisfy himself that "Mark" has the remarkable structure just described. Almost the whole of this Gospel consists of the first component; namely, the *threefold tradition*. But in chap. i. 23-28 he will discover an exorcistic story, not to be found in "Matthew," but repeated, often word for word, in "Luke." This, therefore, belongs to one of the *twofold traditions*. In chap. viii. 1-10, on the other hand, there is a detailed account of the miracle of feeding the four thousand; which is closely repeated in "Matthew" xv. 32-39, but is not to be found in "Luke." This is an example of the other *twofold tradition*, possible in "Mark." Finally, the story of the blind man of Bethsaida, "Mark" viii. 22-26, is *peculiar* to "Mark."

VI. Suppose that, A standing for the *threefold tradition*, or the matter common to all three Gospels; we call the matter common to "Mark" and "Matthew" only—B; that common to "Mark"

and " Luke " only—C ; that common to " Matthew "
and " Luke " only—D ; while the peculiar components of " Mark," " Matthew," and " Luke " are
severally indicated by E, F, G ; then the structure
of the Gospels may be represented thus:

$$\text{Components of "Mark"} = A + B + C + E.$$
$$\text{,,} \quad \text{"Matthew"} = A + B + D + F.$$
$$\text{,,} \quad \text{"Luke"} = A + C + D + G.$$

VII. The analysis of the Synoptic documents
need be carried no further than this point, in
order to suggest one extremely important, and,
apparently unavoidable conclusion ; and that is,
that their authors were neither three independent
witnesses of the things narrated ; nor, for the
parts of the narrative about which all agree, that
is to say, the *threefold tradition*, did they employ
independent sources of information.   It is simply incredible that each of three independent
witnesses of any series of occurrences should
tell a story so similar, not only in arrangement
and in small details, but in words, to that of
each of the others.

Hence it follows, either that the Synoptic
writers have, mediately or immediately, copied
one from the other : or that the three have drawn
from a common source ; that is to say, from one
arrangement of similar traditions (whether oral
or written) ; though that arrangement may have

been extant in three or more, somewhat different versions.

VIII. The suppositions (*a*) that "Mark" had "Matthew" and "Luke" before him; and (*b*) that either of the two latter was acquainted with the work of the other, would seem to involve some singular consequences.

a. The second Gospel is saturated with the lowest supernaturalism. . Jesus is exhibited as a wonder-worker and exorcist of the first rank. The earliest public recognition of the Messiahship of Jesus comes from an "unclean spirit"; he himself is made to testify to the occurrence of the miraculous feeding twice over.

The purpose with which "Mark" sets out is to show forth Jesus as the Son of God, and it is suggested, if not distinctly stated, that he acquired this character at his baptism by John. The absence of any reference to the miraculous events of the infancy, detailed by "Matthew" and "Luke;" or to the appearances after the discovery of the emptiness of the tomb; is unintelligible, if "Mark" knew anything about them, or believed in the miraculous conception. The second Gospel is no summary: "Mark" can find room for the detailed story, irrelevant to his main purpose, of the beheading of John the Baptist, and his miraculous narrations are crowded with minute particulars. Is it to be imagined that, with the supposed apostolic authority of Matthew

before him, he could leave out the miraculous
conception of Jesus and the ascension? Further,
ecclesiastical tradition would have us believe that
Mark wrote down his recollections of what Peter
taught. Did Peter then omit to mention these
matters? Did the fact testified by the oldest
authority extant, that the first appearance of the
risen Jesus was to himself seem not worth men-
tioning? Did he really fail to speak of the great
position in the Church solemnly assigned to him by
Jesus? The alternative would seem to be the
impeachment either of Mark's memory, or of his
judgment. But Mark's memory, is so good that
he can recollect how, on the occasion of the stilling
of the waves, Jesus was asleep " on the cushion,"
he remembers that the woman with the issue had
" spent all she had " on her physicians; that
there was not room " even about the door " on a
certain occasion at Capernaum. And it is surely
hard to believe that " Mark " should have failed to
recollect occurrences of infinitely greater moment,
or that he should have deliberately left them out,
as things not worthy of mention.

b. The supposition that " Matthew " was
acquainted with " Luke," or " Luke " with
" Matthew " has equally grave implications. If
that be so, the one who used the other could have
had but a poor opinion of his predecessor's historical veracity. If, as most experts agree, " Luke "
is later than " Matthew," it is clear that he does

not credit "Matthew's" account of the infancy;
does not believe the "Sermon on the Mount"
as given by Matthew was preached; does not believe in the two feeding miracles, to which Jesus
himself is made to refer; wholly discredits
"Matthew's" account of the events after the
crucifixion; and thinks it not worth while to
notice "Matthew's" grave admission that "some
doubted."

IX. None of these troublesome consequences
pursue the hypothesis that the *threefold tradition*,
in one, or more, Greek versions, was extant before
either of the canonical Synoptic Gospels; and that
it furnished the fundamental framework of their
several narratives. Where and when the threefold narrative arose, there is no positive evidence;
though it is obviously probable that the traditions
it embodies, and perhaps many others, took their
rise in Palestine and spread thence to Asia Minor,
Greece, Egypt and Italy, in the track of the early
missionaries. Nor is it less likely that they
formed part of the "didaskalia" of the primitive
Nazarene and Christian communities.[1]

X. The interest which attaches to "Mark"
arises from the fact that it seems to present this

---

[1] Those who regard the Apocalyptic discourse as a "vaticina-
tion after the event" may draw conclusions therefrom as to the
date of the Gospels in which its several forms occur. But the
assumption is surely dangerous, from an apologetic point of
view, since it begs the question as to the unhistorical character
of this solemn prophecy.

early, probably earliest, Greek Gospel narrative, with least addition, or modification. If, as appears likely from some internal evidences, it was compiled for the use of the Christian sodalities in Rome; and that it was accepted by them as an adequate account of the life and work of Jesus, it is evidence of the most valuable kind respecting their beliefs and the limits of dogma, as conceived by them.

In such case, a good Roman Christian of that epoch might know nothing of the doctrine of the incarnation, as taught by " Matthew " and " Luke "; still less of the "logos" doctrine of "John ", neither need he have believed anything more than the simple fact of the resurrection. It was open to him to believe it either corporeal, or spiritual. He would never have heard of the power of the keys bestowed upon Peter ; nor have had brought to his mind so much as a suggestion of trinitarian doctrine. He might be a rigidly monotheistic Judæo-Christian, and consider himself bound by the law : he might be a Gentile Pauline convert, neither knowing of nor caring for such restrictions. In neither case would he find in "Mark" any serious stumbling-block. In fact, persons of all the categories admitted to salvation by Justin, in the middle of the second century,[1] could accept " Mark " from beginning to end. It may well be, that, in this wide adaptability, backed by the

[1] See p. 287 of this volume

authority of the metropolitan church, there lies
the reason for the fact of the preservation of
" Mark," notwithstanding its limited and dogma·
tically colourless character, as compared with the
Gospels of " Luke " and " Matthew."

XI. " Mark," as we have seen, contains a relatively small body of ethical and religious instruction and only a few parables. Were these
all that existed in the primitive threefold tradition ? Were none others current in the Roman
communities, at the time "Mark" wrote, supposing
he wrote in Rome ? Or, on the other hand, was
there extant, as early as the time at which
" Mark " composed his Greek edition of the
primitive Evangel, one or more collections of
parables and teachings, such as those which form
the bulk of the twofold tradition, common exclusively to " Matthew " and " Luke," and are
also found in their single traditions ? Many have
assumed this, or these, collections to be identical
with, or at any rate based upon, the "logia," of
which ecclesiastical tradition says, that they were
written in Aramaic by Matthew, and that everybody translated them as he could.

Here is the old difficulty again. If such materials were known to " Mark," what imaginable
reason could he have for not using them ? Surely
displacement of the long episode of John the Baptist—even perhaps of the story of the Gadarene
swine—by portions of the Sermon on the Mount or

by one or two of the beautiful parables in the twofold and single traditions would have been great improvements; and might have been effected, even though "Mark" was as much pressed for space as some have imagined. But there is no ground for that imagination; Mark has actually found room for four or five parables; why should he not have given the best, if he had known of them? Admitting he was the mere *pedissequus et breviator* of Matthew, that even Augustine supposed him to be, what could induce him to omit the Lord's Prayer?

Whether more or less of the materials of the twofold tradition D, and of the peculiar traditions F and G, were or were not current in some of the communities, as early as, or perhaps earlier than, the triple tradition, it is not necessary for me to discuss; nor to consider those solutions of the Synoptic problem which assume that it existed earlier, and was already combined with more or less narrative. Those who are working out the final solution of the Synoptic problem are taking into account, more than hitherto, the possibility that the widely separated Christian communities of Palestine, Asia Minor, Egypt, and Italy, especially after the Jewish war of A.D. 66-70, may have found themselves in possession of very different traditional materials. Many circumstances tend to the conclusion that, in Asia Minor, even the narrative part of the threefold tradition had a formidable rival; and that, around this second narrative, teaching traditions of a totally different order from those in the Synoptics, grouped themselves; and, under the influence of converts imbued more or less with the philosophical speculations of the time, eventually took shape in the fourth Gospel and its associated literature.

XII. But it is unnecessary, and it would be out of place, for me to attempt to do more than indicate the existence of these complex and difficult questions. My purpose has been to make it clear that the Synoptic problem must force itself upon every one who studies the Gospels with attention; that the broad facts of the case, and some of the consequences deducible from these facts, are just as plain to the simple English reader as they are to the profoundest scholar.

One of these consequences is that the three-fold tradition presents us with a narrative believed to be historically true, in all particulars, by the major part, if not the whole, of the Christian communities. That narrative is penetrated, from beginning to end, by the demonological beliefs of which the Gadarene story is a specimen; and, if the fourth Gospel indicates the existence of another and, in some respects, irreconcilably divergent narrative, in which the demonology retires into the background, it is none the less there.

Therefore, the demonology is an integral and inseparable component of primitive Christianity.

The farther back the origin of the gospels is dated, the stronger does the certainty of this conclusion grow; and the more difficult it becomes to suppose that Jesus himself may not have shared the superstitious beliefs of his disciples.

It further follows that those who accept devils, possession, and exorcism as essential elements of their conception of the spiritual world may consistently consider the testimony of the Gospels to be unimpeachable in respect of the information they give us respecting other matters which appertain to that world.

Those who reject the gospel demonology, on the other hand, would seem to be as completely barred, as I feel myself to be, from professing to take the accuracy of that information for granted. If the threefold tradition is wrong about one fundamental topic, it may be wrong about another, while the authority of the single traditions, often mutually contradictory as they are, becomes a vanishing quantity.

It really is unreasonable to ask any rejector of the demonology to say more with respect to those other matters, than that the statements regarding them may be true, or may be false; and that the ultimate decision, if it is to be favourable, must depend on the production of testimony of a very different character from that of the writers of the four gospels. Until such evidence is brought forward, that refusal of assent, with willingness to re-open the question, on cause shown, which is what I mean by Agnosticism, is, for me the only course open.

A verdict of "not proven" is undoubtedly unsatisfactory and essentially provisional, so far forth as the subject of the trial is capable of being dealt with by due process of reason.

Those who are of opinion that the historical realities at the root of Christianity, lie beyond the jurisdiction of science, need not be considered. Those who are convinced that the evidence is, and must always remain, insufficient to support any definite conclusion, are justified in ignoring the subject. They must be content to put up with that reproach of being mere destroyers, of which Strauss speaks. They may say that there are so many problems which are and must remain insoluble, that the "burden of the mystery" "of all this unintelligible world" is not appreciably affected by one more or less.

For myself, I must confess that the problem of the origin of such very remarkable historical phenomena as the doctrines, and the social organization, which, in their broad features certainly existed, and were in a state of rapid development, within a hundred years of the crucifixion of Jesus; and which have steadily prevailed against all rivals, among the most intelligent and civilized nations in the world ever since,

is, and always has been, profoundly interesting;
and, considering how recent the really scientific
study of that problem, and how great the progress
made during the last half century in supplying the
conditions for a positive solution of the problem,
I cannot doubt that the attainment of such a
solution is a mere question of time.

I am well aware that it has lain far beyond my
powers to take any share in this great undertaking. All that I can hope is to have done
somewhat towards "the preparation of those who
have ceased to be contented with the old and find
no satisfaction in half measures": perhaps, also,
something towards the lessening of that great
proportion of my countrymen, whose eminent
characteristic it is that they find full "full satisfaction in half measures."   T. H. H.

HODESLEA, EASTBOURNE,
   December 4th, 1893.

CONTENTS

c

VI

VII

VIII

IX

X

XI

COLLECTED ESSAYS

VOLUME V

I

PROLOGUE

[*Controverted Questions*, 1892]

Le plus grand service qu'on puisse rendre à la science est d'y faire place nette avant d'y rien construire.—CUVIER.

MOST of the Essays comprised in the present volume have been written during the last six or seven years, without premeditated purpose or intentional connection, in reply to attacks upon doctrines which I hold to be well founded; or in refutation of allegations respecting matters lying within the province of natural knowledge, which I believe to be erroneous; and they bear the mark of their origin in the controversial tone which pervades them.

Of polemical writing, as of other kinds of warfare, I think it may be said, that it is often useful, sometimes necessary, and always more or less of an evil. It is useful, when it attracts attention to topics which might otherwise be neglected; and when, as does sometimes happen, those who come to see a contest remain to think. It is necessary,

when the interests of truth and of justice are at
stake. It is an evil, in so far as controversy
always tends to degenerate into quarrelling, to
swerve from the great issue of what is right and
what is wrong to the very small question of who
is right and who is wrong. I venture to hope
that the useful and the necessary were more
conspicuous than the evil attributes of literary
militancy, when these papers were first published;
but I have had some hesitation about reprinting
them. If I may judge by my own taste, few
literary dishes are less appetising than cold
controversy; moreover, there is an air of unfairness about the presentation of only one side of
a discussion, and a flavour of unkindness in the
reproduction of "winged words," which, however
appropriate at the time of their utterance, would
find a still more appropriate place in oblivion.
Yet, since I could hardly ask those who have
honoured me by their polemical attentions to
confer lustre on this collection, by permitting me
to present their lucubrations along with my own;
and since it would be a manifest wrong to them to
deprive their, by no means rare, vivacities of
language of such justification as they may derive
from similar freedoms on my part; I came to the
conclusion that my best course was to leave the
essays just as they were written;[1] assuring my

[1] With a few exceptions, which are duly noted when they
amount to more than verbal corrections.

honourable adversaries than any heat of which
signs may remain was generated, in accordance
with the law of the conservation of energy, by the
force of their own blows, and has long since been
dissipated into space.

But, however the polemical concomitants of
these discussions may be regarded—or better, disregarded—there is no doubt either about the importance of the topics of which they treat, or
as to the public interest in the "Controverted
Questions" with which they deal. Or rather,
the Controverted Question; for disconnected as
these pieces may, perhaps, appear to be, they are,
in fact, concerned only with different aspects of a
single problem, with which thinking men have
been occupied, ever since they began seriously to
consider the wonderful frame of things in which
their lives are set, and to seek for trustworthy
guidance among its intricacies.

Experience speedily taught them that the
shifting scenes of the world's stage have a permanent background; that there is order amidst the
seeming confusion, and that many events take
place according to unchanging rules. To this
region of familiar steadiness and customary regu-
larity they gave the name of Nature. But at the
same time, their infantile and untutored reason,
little more, as yet, than the playfellow of the
imagination, led them to believe that this tangible,
commonplace, orderly world of Nature was sur-

rounded and interpenetrated by another intangible and mysterious world, no more bound by fixed rules than, as they fancied, were the thoughts and passions which coursed through their minds and seemed to exercise an intermittent and capricious rule over their bodies. They attributed to the entities, with which they peopled this dim and dreadful region, an unlimited amount of that power of modifying the course of events of which they themselves possessed a small share, and thus came to regard them as not merely beyond, but above, Nature.

Hence arose the conception of a " Supernature " antithetic to " Nature "—the primitive dualism of a natural world " fixed in fate " and a super-natural, left to the free play of volition—which has pervaded all later speculation and, for thousands of years, has exercised a profound influence on practice. For it is obvious that, on this theory of the Universe, the successful conduct of life must demand careful attention to both worlds; and, if either is to be neglected, it may be safer that it should be Nature. In any given contingency, it must doubtless be desirable to know what may be expected to happen in the ordinary course of things; but it must be quite as necessary to have some inkling of the line likely to be taken by supernatural agencies able, and possibly willing, to suspend or reverse that course. Indeed, logically developed, the dualistic theory must needs end in almost exclusive attention to Supernature, and in trust that its over-ruling strength will be exerted in favour of those who stand well with its denizens. On the other hand, the lessons of the great schoolmaster, experience, have hardly seemed to accord with this conclusion. They have taught, with considerable emphasis, that it does not answer to neglect Nature; and that, on the whole, the more attention paid to her dictates the better men fare.

Thus the theoretical antithesis brought about a practical antagonism. From the earliest times of which we have any knowledge, Naturalism· and Supernaturalism have consciously, or unconsciously, competed and struggled with one another; and the varying fortunes of the contest are written in the records of the course of civilisation, from those of Egypt and Babylonia, six thousand years ago, down to those of our own time and people.

These records inform us that, so far as men have paid attention to Nature, they have been rewarded for their pains. They have developed the Arts which have furnished the conditions of civilised existence; and the Sciences, which have been a progressive revelation of reality and have afforded the best discipline of the mind in the methods of discovering truth. They have accumulated a vast body of universally accepted knowledge; and the conceptions of man and of society,

of morals and of law, based upon that knowledge,
are every day more and more, either openly or
tacitly, acknowledged to be the foundations of
right action.

History also tells us that the field of the
supernatural has rewarded its cultivators with a
harvest, perhaps not less luxuriant, but of a
different character. It has produced an almost
infinite diversity of Religions. These, if we set
aside the ethical concomitants upon which natural
knowledge also has a claim, are composed of
information about Supernature; they tell us of
the attributes of supernatural beings, of their
relations with Nature, and of the operations by
which their interference with the ordinary course
of events can be secured or averted. It does
not appear, however, that supernaturalists have
attained to any agreement about these matters, or
that history indicates a widening of the influence
of supernaturalism on practice, with the onward
flow of time. On the contrary, the various
religions are, to a great extent, mutually exclusive; and their adherents delight in charging
each other, not merely with error, but with
criminality, deserving and ensuing punishment
of infinite severity. In singular contrast with
natural knowledge, again, the acquaintance of
mankind with the supernatural appears the more
extensive and the more exact, and the influence
of supernatural doctrines upon conduct the greater,

the further back we go in time and the lower the stage of civilisation submitted to investigation. Historically, indeed, there would seem to be an inverse relation between supernatural and natural knowledge. As the latter has widened, gained in precision and in trustworthiness, so has the former shrunk, grown vague and questionable; as the one has more and more filled the sphere of action, so has the other retreated into the region of meditation, or vanished behind the screen of mere verbal recognition.

Whether this difference of the fortunes of Naturalism and of Supernaturalism is an indication of the progress, or of the regress, of humanity; of a fall from, or an advance towards, the higher life; is a matter of opinion. The point to which I wish to direct attention is that the difference exists and is making itself felt. Men are growing to be seriously alive to the fact that the historical evolution of humanity, which is generally, and I venture to think not unreasonably, regarded as progress, has been, and is being, accompanied by a co-ordinate elimination of the supernatural from its originally large occupation of men's thoughts. The question—How far is this process to go?—is, in my apprehension, the Controverted Question of our time.

Controversy on this matter—prolonged, bitter, and fought out with the weapons of the flesh, as

well as with those of the spirit—is no new thing
to Englishmen.   We have been more or less
occupied with it these five hundred years.   And,
during that time, we have made attempts to
establish a *modus vivendi* between the antagonists,
some of which have had a world-wide influence ;
though, unfortunately, none have proved universally and permanently satisfactory.

In the fourteenth century, the controverted
question among us was, whether certain portions
of the Supernaturalism of mediæval Christianity
were well-founded.   John Wicliff proposed a
solution of the problem which, in the course of
the following two hundred years, acquired wide
popularity and vast historical importance : Lollards,
Hussites, Lutherans, Calvinists, Zwinglians, Socin-
ians, and Anabaptists, whatever their disagreements, concurred in the proposal to reduce the
Supernaturalism of Christianity within the limits
sanctioned by the Scriptures.   None of the chiefs
of Protestantism called in question either the
supernatural origin and infallible authority of the
Bible, or the exactitude of the account of the
supernatural world given in its pages.   In fact,
they could not afford to entertain any doubt
about these points, since the infallible Bible was
the fulcrum of the lever with which they were
endeavouring to upset the Chair of St. Peter.
The "freedom of private judgment" which they
proclaimed, meant no more, in practice, than

permission to themselves to make free with the
public judgment of the Roman Church, in respect
of the canon and of the meaning to be attached
to the words of the canonical books. Private
judgment—that is to say, reason—was (theoretically, at any rate) at liberty to decide what books
were and what were not to take the rank of
"Scripture"; and to determine the sense of any
passage in such books. But this sense, once
ascertained to the mind of the sectary, was to be
taken for pure truth—for the very word of God.
The controversial efficiency of the principle of
biblical infallibility lay in the fact that the con-
servative adversaries of the Reformers were not in
a position to contravene it without entangling
themselves in serious difficulties; while, since
both Papists and Protestants agreed in taking
efficient measures to stop the mouths of any more
radical critics, these did not count.

The impotence of their adversaries, however, did
not remove the inherent weakness of the position
of the Protestants. The dogma of the infallibility
of the Bible is no more self-evident than is that
of the infallibility of the Pope. If the former is
held by "faith," then the latter may be. If the
latter is to be accepted, or rejected, by private
judgment, why not the former? Even if the
Bible could be proved anywhere to assert its own
infallibility, the value of that self-assertion to
those who dispute the point is not obvious. On

it followed that, in the long run, whoso settled the canon defined the creed. If the private judgment of Luther might legitimately conclude that the epistle of James was contemptible, while the epistles of Paul contained the very essence of Christianity, it must be permissible for some other private judgment, on as good or as bad grounds, to reverse these conclusions; the critical process which excluded the Apocrypha could not be barred, at any rate by people who rejected the authority of the Church, from extending its operations to Daniel, the Canticles, and Ecclesiastes; nor, having got so far, was it easy to allege any good ground for staying the further progress of criticism. In fact, the logical development of Protestantism could not fail to lay the authority of the Scriptures at the feet of Reason; and, in the hands of latitudinarian and rationalistic theologians, the despotism of the Bible was rapidly converted into an extremely limited monarchy. Treated with as much respect as ever, the sphere of its practical authority was minimised; and its decrees were valid only so far as they were countersigned by common sense, the responsible minister.

The champions of Protestantism are much given to glorify the Reformation of the sixteenth century as the emancipation of Reason; but it may be doubted if their contention has any solid ground; while there is a good deal of evidence to

show, that aspirations after intellectual freedom
had nothing whatever to do with the movement.
Dante, who struck the Papacy as hard blows as
Wicliff; Wicliff himself and Luther himself, when
they began their work ; were far enough from
any intention of meddling with even the most
irrational of the dogmas of mediæval Supernaturalism. From Wicliff to Socinus, or even to
Münzer, Rothmann, and John of Leyden, I fail to
find a trace of any desire to set reason free. The
most that can be discovered is a proposal to
change masters. From being the slave of the
Papacy the intellect was to become the serf of the
Bible ; or, to speak more accurately, of somebody's
interpretation of the Bible, which, rapidly shifting
its attitude from the humility of a private judgment to the arrogant Cæsaro-papistry of a state-
enforced creed, had no more hesitation about
forcibly extinguishing opponent private judgments
and judges, than had the old-fashioned Pontiff-
papistry.

It was the iniquities, and not the irrationalities,
of the Papal system that lay at the bottom of the
revolt of the laity ; which was, essentially, an
attempt to shake off the intolerable burden of
certain practical deductions from a Supernaturalism in which everybody, in principle, acquiesced.
What was the gain to intellectual freedom of
abolishing transubstantiation, image worship, in-
dulgences, ecclesiastical infallibility; if consub-

stantiation, real-unreal presence mystifications, the bibliolatry, the "inner-light" pretensions, and the demonology, which are fruits of the same supernaturalistic tree, remained in enjoyment of the spiritual and temporal support of a new infallibility? One does not free a prisoner by merely scraping away the rust from his shackles.

It will be asked, perhaps, was not the Reformation one of the products of that great outbreak of many-sided free mental activity included under the general head of the Renascence? Melanchthon, Ulrich von Hutten, Beza, were they not all humanists? Was not the arch-humanist, Erasmus, fautor-in-chief of the Reformation, until he got frightened and basely deserted it?

From the language of Protestant historians, it would seem that they often forget that Reformation and Protestantism are by no means convertible terms. There were plenty of sincere and indeed zealous reformers, before, during, and after the birth and growth of Protestantism, who would have nothing to do with it. Assuredly, the rejuvenescence of science and of art; the widening of the field of Nature by geographical and astronomical discovery; the revelation of the noble ideals of antique literature by the revival of classical learning; the stir of thought, throughout all classes of society, by the printers' work, loosened traditional bonds and weakened the hold of mediæval Supernaturalism. In the interests of liberal culture and of national welfare, the humanists were eager to lend a hand to anything which tended to the discomfiture of their sworn enemies, the monks, and they willingly supported every movement in the direction of weakening ecclesiastical interference with civil life. But the bond of a common enemy was the only real tie between the humanist and the protestant; their alliance was bound to be of short duration, and, sooner or later, to be replaced by internecine warfare. The goal of the humanists, whether they were aware of it or not, was the attainment of the complete intellectual freedom of the antique philosopher, than which nothing could be more abhorrent to a Luther, a Calvin, a Beza, or a Zwingli.

The key to the comprehension of the conduct of Erasmus, seems to me to lie in the clear apprehension of this fact. That he was a man of many weaknesses may be true; in fact, he was quite aware of them and professed himself no hero. But he never deserted that reformatory movement which he originally contemplated; and it was impossible he should have deserted the specifically Protestant reformation in which he never took part. He was essentially a theological whig, to whom radicalism was as hateful as it is to all whigs; or, to borrow a still more appropriate comparison from modern times, a broad churchman who refused to enlist with either the High

Church or the Low Church zealots, and paid the penalty of being called coward, time-server and traitor, by both. Yet really there is a good deal in his pathetic remonstrance that he does not see why he is bound to become a martyr for that in which he does not believe ; and a fair consideration of the circumstances and the consequences of the Protestant reformation seems to me to go a long way towards justifying the course he adopted.

Few men had better means of being acquainted with the condition of Europe ; none could be more competent to gauge the intellectual shallowness and self-contradiction of the Protestant criticism of Catholic doctrine ; and to estimate, at its proper value, the fond imagination that the waters let out by the Renascence would come to rest amidst the blind alleys of the new ecclesiasticism. The bastard, whilom poor student and monk, become the familiar of bishops and princes, at home in all grades of society, could not fail to be aware of the gravity of the social position, of the dangers imminent from the profligacy and indifference of the ruling classes, no less than from the anarchical tendencies of the people who groaned under their oppression. The wanderer who had lived in Germany, in France, in England, in Italy, and who counted many of the best and most influential men in each country among his friends, was not likely to estimate wrongly the enormous forces which were still at the command of the

istic clericalism in Geneva and in Scotland; the
long agony of religious wars, persecutions, and
massacres, which devastated France and reduced
Germany almost to savagery; finishing with the
spectacle of Lutheranism in its native country
sunk into mere dead Erastian formalism, before
it was a century old; while Jesuitry triumphed
over Protestantism in three-fourths of Europe,
bringing in its train a recrudescence of all the
corruptions Erasmus and his friends sought to
abolish; might not he have quite honestly
thought this a somewhat too heavy price to pay
for Protestantism; more especially, since no one
was in a better position than himself to know
how little the dogmatic foundation of the new
confessions was able to bear the light which the
inevitable progress of humanistic criticism would
throw upon them?   As the wiser of his contemporaries saw, Erasmus was, at heart, neither
Protestant nor Papist, but an "Independent
Christian"; and, as the wiser of his modern
biographers have discerned, he was the precursor,
not of sixteenth century reform, but of eighteenth
century "enlightenment"; a sort of broad-church
Voltaire, who held by his "Independent Christianity" as stoutly as Voltaire by his Deism.

In fact, the stream of the Renascence, which
bore Erasmus along, left Protestantism stranded
amidst the mudbanks of its articles and creeds:
while its true course became visible to all men,

two centuries later. By this time, those in whom
the movement of the Renascence was incarnate
became aware what spirit they were of; and they
attacked Supernaturalism in its Biblical strong-
hold, defended by Protestants and Romanists
with equal zeal. In the eyes of the " Patriarch,"
Ultramontanism, Jansenism, and Calvinism were
merely three persons of the one " Infâme " which
it was the object of his life to crush. If he
hated one more than another, it was probably the
last; while D'Holbach, and the extreme left of
the free-thinking host, were disposed to show no
more mercy to Deism and Pantheism.

The sceptical insurrection of the eighteenth
century made a terrific noise and frightened not
a few worthy people out of their wits; but cool
judges might have foreseen, at the outset, that
the efforts of the later rebels were no more likely
than those of the earlier, to furnish permanent
resting-places for the spirit of scientific inquiry.
However worthy of admiration may be the acuteness, the common sense, the wit, the broad
humanity, which abound in the writings of the
best of the free-thinkers; there is rarely much to
be said for their work as an example of the
adequate treatment of a grave and difficult investigation. I do not think any impartial judge
will assert that, from this point of view, they are
much better than their adversaries. It must be
admitted that they share to the full the fatal

weakness of *a priori* philosophising, no less than
the moral frivolity common to their age ; while a
singular want of appreciation of history, as the
record of the moral and social evolution of the
human race, permitted them to resort to pre-
posterous theories of imposture, in order to
account for the religious phenomena which are
natural products of that evolution.

For the most part, the Romanist and Protestant
adversaries of the free-thinkers met them with
arguments no better than their own ; and with
vituperation, so far inferior that it lacked the wit.
But one great Christian Apologist fairly captured
the guns of the free-thinking array, and turned
their batteries upon themselves. Speculative
" infidelity" of the eighteenth century type was
mortally wounded by the *Analogy ;* while the progress of the historical and psychological sciences
brought to light the important part played by the
mythopœic faculty ; and, by demonstrating the
extreme readiness of men to impose upon themselves, rendered the calling in of sacerdotal
cooperation, in most cases, a superfluity.

Again, as in the fourteenth and the sixteenth
centuries, social and political influences came into
play. The free-thinking *philosophes,* who objected
to Rousseau's sentimental religiosity almost as
much as they did to *L'Infâme,* were credited with
the responsibility for all the evil deeds of
Rousseau's Jacobin disciples, with about as much

justification as Wicliff was held responsible for the Peasants' revolt, or Luther for the *Bauern-krieg*. In England, though our *ancien régime* was not altogether lovely, the social edifice was never in such a bad way as in France; it was still capable of being repaired; and our forefathers, very wisely, preferred to wait until that operation could be safely performed, rather than pull it all down about their ears, in order to build a philosophically planned house on brand-new speculative foundations. Under these circumstances, it is not wonderful that, in this country, practical men preferred the gospel of Wesley and Whitfield to that of Jean Jacques; while enough of the old leaven of Puritanism remained to ensure the favour and support of a large number of religious men to a revival of evangelical supernaturalism. Thus, by degrees, the free-thinking, or the indifference, prevalent among us in the first half of the eighteenth century, was replaced by a strong supernaturalistic reaction, which submerged the work of the free-thinkers; and even seemed, for a time, to have arrested the naturalistic movement of which that work was an imperfect indication. Yet, like Lollardry, four centuries earlier, free-thought merely took to running underground, safe, sooner or later, to return to the surface.

My memory, unfortunately, carries me back to the fourth decade of the nineteenth century, when the

evangelical flood had a little abated and the tops
of certain mountains were soon to appear, chiefly
in the neighbourhood of Oxford; but when nevertheless, bibliolatry was rampant; when church
and chapel alike proclaimed, as the oracles of God,
the crude assumptions of the worst informed and,
in natural sequence, the most presumptuously
bigoted, of all theological schools.

In accordance with promises made on my
behalf, but certainly without my authorisation, I
was very early taken to hear "sermons in the
vulgar tongue." And vulgar enough often was
the tongue in which some preacher, ignorant alike
of literature, of history, of science, and even of
theology, outside that patronised by his own
narrow school, poured forth, from the safe
entrenchment of the pulpit, invectives against
those who deviated from his notion of orthodoxy.
From dark allusions to "sceptics" and "infidels,"
I became aware of the existence of people who
trusted in carnal reason; who audaciously doubted
that the world was made in six natural days, or
that the deluge was universal; perhaps even went
so far as to question the literal accuracy of the
story of Eve's temptation, or of Balaam's ass; and,
from the horror of the tones in which they were
mentioned, I should have been justified in drawing
the conclusion that these rash men belonged to the
criminal classes. At the same time, those who
were more directly responsible for providing me

with the knowledge essential to the right
guidance of life (and who sincerely desired to do
so), imagined they were discharging that most
sacred duty by impressing upon my childish mind
the necessity, on pain of reprobation in this world
and damnation in the next, of accepting, in the
strict and literal sense, every statement contained
in the Protestant Bible. I was told to believe,
and I did believe, that doubt about any of them
was a sin, not less reprehensible than a moral
delict. I suppose that, out of a thousand of my
contemporaries, nine hundred, at least, had their
minds systematically warped and poisoned, in the
name of the God of truth, by like discipline. I am
sure that, even a score of years later, those who
ventured to question the exact historical accuracy
of any part of the Old Testament and *a fortiori* of
the Gospels, had to expect a pitiless shower of
verbal missiles, to say nothing of the other dis-
agreeable consequences which visit those who, in
any way, run counter to that chaos of prejudices
called public opinion.

My recollections of this time have recently been
revived by the perusal of a remarkable document,[1]
signed by as many as thirty-eight out of the
twenty odd thousand clergymen of the Established
Church. It does not appear that the signataries
are officially accredited spokesmen of the ecclesias-

---

[1] *Declaration on the Truth of Holy Scripture.* The *Times,*
18th December, 1891.

tical corporation to which they belong; but I feel
bound to take their word for it, that they are
"stewards of the Lord, who have received the Holy
Ghost," and, therefore, to accept this memorial as
evidence that, though the Evangelicism of my early
days may be deposed from its place of power,
though so many of the colleagues of the thirty-eight
even repudiate the title of Protestants, yet the
green bay tree of bibliolatry flourishes as it did sixty
years ago. And, as in those good old times, whoso
refuses to offer incense to the idol is held to be guilty
of "a dishonour to God," imperilling his salvation.

It is to the credit of the perspicacity of the
memorialists that they discern the real nature of
the Controverted Question of the age. They are
awake to the unquestionable fact that, if Scripture
has been discovered "not to be worthy of un-
questioning belief," faith "in the supernatural
itself" is, so far, undermined. And I may con-
gratulate myself upon such weighty confirmation
of an opinion in which I have had the fortune to
anticipate them. But whether it is more to the
credit of the courage, than to the intelligence, of
the thirty-eight that they should go on to pro-
claim that the canonical scriptures of the Old
and New Testaments "declare incontrovertibly
the actual historical truth in all records, both of
past events and of the delivery of predictions to
be thereafter fulfilled," must be left to the coming
generation to decide.

The interest which attaches to this singular document will, I think, be based by most thinking men, not upon what it is, but upon that of which it is a sign. It is an open secret, that the memorial is put forth as a counterblast to a manifestation of opinion of a contrary character, on the part of certain members of the same ecclesiastical body, who therefore have, as I suppose, an equal right to declare themselves "stewards of the Lord and recipients of the Holy Ghost." In fact, the stream of tendency towards Naturalism, the course of which I have briefly traced, has, of late years, flowed so strongly, that even the Churches have begun, I dare not say to drift, but, at any rate, to swing at their moorings. Within the pale of the Anglican establishment, I venture to doubt, whether, at this moment, there are as many thorough-going defenders of "plenary inspiration" as there were timid questioners of that doctrine, half a century ago.   Commentaries, sanctioned by the highest authority, give up the "actual historical truth" of the cosmogonical and diluvial narratives.   University professors of deservedly high repute accept the critical decision that the Hexateuch is a compilation, in which the share of Moses, either as author or as editor, is not quite so clearly demonstrable as it might be ; highly placed Divines tell us that the pre-Abrahamic Scripture narratives may be ignored ; that the book of Daniel may be regarded as a

patriotic romance of the second century B.C.;
that the words of the writer of the fourth Gospel
are not always to be distinguished from those
which he puts into the mouth of Jesus. Conser-
vative, but conscientious, revisers decide that
whole passages, some of dogmatic and some of
ethical importance, are interpolations. An uneasy
sense of the weakness of the dogma of Biblical
infallibility seems to be at the bottom of a
prevailing tendency once more to substitute the
authority of the " Church " for that of the Bible.
In my old age, it has happened to me to be taken
to task for regarding Christianity as a "religion
of a book" as gravely as, in my youth, I should
have been reprehended for doubting that proposition. It is a no less interesting symptom that
the State Church seems more and more anxious
to repudiate all complicity with the principles of
the Protestant Reformation and to call itself
" Anglo-Catholic." Inspiration, deprived of its
old intelligible sense, is watered down into a
mystification. The Scriptures are, indeed, inspired; but they contain a wholly undefined and
indefinable " human element "; and this unfortunate intruder is converted into a sort of biblical
whipping boy. Whatsoever scientific investigation,
historical or physical, proves to be erroneous, the
" human element " bears the blame; while the
divine inspiration of such statements, as by their
nature are out of reach of proof or disproof, is

Grant that it is "the traditionary testimony of the Church" which guarantees the canonicity of each and all of the books of the Old and New Testaments. Grant also that canonicity means infallibility; yet, according to the thirty-eight, this "traditionary testimony" has to be "ascertained and verified by appeal to antiquity." But "ascertainment and verification" are purely intellectual processes, which must be conducted according to the strict rules of scientific investigation, or be self-convicted of worthlessness. Moreover, before we can set about the appeal to "antiquity," the exact sense of that usefully vague term must be defined by similar means. "Antiquity" may include any number of centuries, great or small; and whether "antiquity" is to comprise the Council of Trent, or to stop a little beyond that of Nicæa, or to come to an end in the time of Irenæus, or in that of Justin Martyr, are knotty questions which can be decided, if at all, only by those critical methods which the signataries treat so cavalierly. And yet the decision of these questions is fundamental, for as the limits of the canonical scriptures vary, so may the dogmas deduced from them require modification. Christianity is one thing, if the fourth Gospel, the Epistle to the Hebrews, the pastoral Epistles, and the Apocalypse are canonical and (by the hypothesis) infallibly true; and another thing, if they are not.

that they should show cause why, in these
days, science should not resume the work the
ancients did so imperfectly, and carry it out
efficiently.

But no such cause can be shown. If "antiquity"
permitted Eusebius, Origen, Tertullian, Irenæus,
to argue for the reception of this book into the
canon and the rejection of that, upon rational
grounds, "antiquity" admitted the whole principle of modern criticism. If Irenæus produces
ridiculous reasons for limiting the Gospels to four,
it was open to any one else to produce good
reasons (if he had them) for cutting them down
to three, or increasing them to five. If the
Eastern branch of the Church had a right to
reject the Apocalypse and accept the Epistle to
the Hebrews, and the Western an equal right to
accept the Apocalypse and reject the Epistle,
down to the fourth century, any other branch
would have an equal right, on cause shown, to
reject both, or, as the Catholic Church afterwards
actually did, to accept both.

Thus I cannot but think that the thirty-eight
are hoist with their own petard. Their "appeal to
antiquity" turns out to be nothing but a round-
about way of appealing to the tribunal, the jurisdiction of which they affect to deny. Having
rested the world of Christian supernaturalism on
the elephant of biblical infallibility, and furnished
the elephant with standing ground on the tortoise

of "antiquity,' they, like their famous Hindoo analogue, have been content to look no further ; and have thereby been spared the horror of discovering that the tortoise rests on a grievously fragile construction, to a great extent the work of that very intellectual operation which they anathematise and repudiate.

Moreover, there is another point to be considered. It is of course true that a Christian Church (whether the Christian Church, or not, depends on the connotation of the definite article) existed before the Christian scriptures ; and that the infallibility of these depends upon the infallibility of the judgment of the persons who selected the books of which they are composed, out of the mass of literature current among the early Christians. The logical acumen of Augustine showed him that the authority of the Gospel he preached must rest on that of the Church to which he belonged.[1] But it is no less true that the Hebrew and the Septuagint versions of most, if not all, of the Old Testament books existed before the birth of Jesus of Nazareth ; and that their divine authority is presupposed by, and therefore can hardly depend upon, the religious body constituted by his disciples. As everybody knows, the very conception of a "Christ" is purely

---

[1] Ego vero evangelio non crederem, nisi ecclesiæ Catholicæ me commoveret auctoritas.—*Contra Epistolam Manichœi*, cap. v.

Jewish. The validity of the argument from the Messianic prophecies vanishes unless their infallible authority is granted; and, as a matter of fact, whether we turn to the Gospels, the Epistles, or the writings of the early Apologists, the Jewish scriptures are recognised as the highest court of appeal of the Christian.

The proposal to cite Christian "antiquity" as a witness to the infallibility of the Old Testament, when its own claims to authority vanish, if certain propositions contained in the Old Testament are erroneous, hardly satisfies the requirements of lay logic. It is as if a claimant to be sole legatee, under another kind of testament, should offer his assertion as sufficient evidence of the validity of the will. And, even were not such a circular, or rather rotatory, argument, that the infallibility of the Bible is testified by the infallible Church, whose infallibility is testified by the infallible Bible, too absurd for serious consideration, it remains permissible to ask, Where and when the Church, during the period of its infallibility, as limited by Anglican dogmatic necessities, has officially decreed the "actual historical truth of all records" in the Old Testament? Was Augustine heretical when he denied the actual historical truth of the record of the Creation? Father Suarez, standing on later Roman tradition, may have a right to declare that he was; but it does not lie in the mouth of those who limit their appeal to that early "antiquity," in which Augustine played so great a part, to say so.

Among the watchers of the course of the world of thought, some view with delight and some with horror, the recrudescence of Supernaturalism which manifests itself among us, in shapes ranged along the whole flight of steps, which, in this case, separates the sublime from the ridiculous—from Neo-Catholicism and Inner-light mysticism, at the top, to unclean things, not worthy of mention in the same breath, at the bottom. In my poor opinion, the importance of these manifestations is often greatly over-estimated. The extant forms of Supernaturalism have deep roots in human nature, and will undoubtedly die hard; but, in these latter days, they have to cope with an enemy whose full strength is only just beginning to be put out, and whose forces, gathering strength year by year, are hemming them round on every side. This enemy is Science, in the acceptation of systematised natural knowledge, which, during the last two centuries, has extended those methods of investigation, the worth of which is confirmed by daily appeal to Nature, to every region in which the Supernatural has hitherto been recognised.

When scientific historical criticism reduced the annals of heroic Greece and of regal Rome to the level of fables; when the unity of authorship of the *Iliad* was successfully assailed by scientific literary criticism; when scientific physical criticism, after exploding the geocentric theory of the universe and reducing the solar system itself to one of millions of groups of like cosmic specks, circling, at unimaginable distances from one another through infinite space, showed the supernaturalistic theories of the duration of the earth and of life upon it, to be as inadequate as those of its relative dimensions and importance had been; it needed no prophetic gift to see that, sooner or later, the Jewish and the early Christian records would be treated in the same manner; that the authorship of the Hexateuch and of the Gospels would be as severely tested; and that the evidence in favour of the veracity of many of the statements found in the Scriptures would have to be strong indeed, if they were to be opposed to the conclusions of physical science. In point of fact, so far as I can discover, no one competent to judge of the evidential strength of these conclusions, ventures now to say that the biblical accounts of the creation and of the deluge are true in the natural sense of the words of the narratives. The most modern Reconcilers venture upon is to affirm, that some quite different sense may be put upon the words; and that this non-natural sense may, with a little trouble, be manipulated into some sort of non-contradiction of scientific truth.

My purpose, in the essay (XVI.) which treats of the narrative of the Deluge, was to prove, by

physical criticism, that no such event as that
described ever took place ; to exhibit the untrustworthy character of the narrative demonstrated
by literary criticism ; and, finally, to account for
its origin, by producing a form of those ancient
legends of pagan Chaldæa, from which the biblical
compilation is manifestly derived. I have yet to
learn that the main propositions of this essay can
be seriously challenged.

In the essays (II., III.) on the narrative of the
Creation, I have endeavoured to controvert the
assertion that modern science supports, either the
interpretation put upon it by Mr. Gladstone, or
any interpretation which is compatible with the
general sense of the narrative, quite apart from
particular details. The first chapter of Genesis
teaches the supernatural creation of the present
forms of life ; modern science teaches that they
have come about by evolution. The first chapter
of Genesis teaches the successive origin—firstly,
of all the plants, secondly, of all the aquatic and
aerial animals, thirdly, of all the terrestrial animals, which now exist—during distinct intervals
of time ; modern science teaches that, throughout
all the duration of an immensely long past, so far
as we have any adequate knowledge of it (that is
as far back as the Silurian epoch), plants, aquatic,
aerial, and terrestrial animals have co-existed ;
that the earliest known are unlike those which at
present exist ; and that the modern species have

come into existence as the last terms of a series,
the members of which have appeared one after
another. Thus, far from confirming the account
in Genesis, the results of modern science, so far as
they go, are in principle, as in detail, hopelessly
discordant with it.

Yet, if the pretensions to infallibility set up,
not by the ancient Hebrew writings themselves,
but by the ecclesiastical champions and friends
from whom they may well pray to be delivered,
thus shatter themselves against the rock of
natural knowledge, in respect of the two most
important of all events, the origin of things and
the palingenesis of terrestrial life, what historical
credit dare any serious thinker attach to the
narratives of the fabrication of Eve, of the Fall,
of the commerce between the *Bene Elohim* and
the daughters of men, which lie between the
creational and the diluvial legends ? And, if
these are to lose all historical worth, what becomes of the infallibility of those who, according
to the later scriptures, have accepted them,
argued from them, and staked far-reaching dogmatic conclusions upon their historical accuracy ?

It is the merest ostrich policy for contemporary
ecclesiasticism to try to hide its Hexateuchal
head—in the hope that the inseparable connection of its body with pre-Abrahamic legends may
be overlooked. The question will still be asked,
if the first nine chapters of the Pentateuch are

unhistorical, how is the historical accuracy of
the remainder to be guaranteed? What more
intrinsic claim has the story of the Exodus than
that of the Deluge, to belief? If God did not
walk in the Garden of Eden, how can we be
assured that he spoke from Sinai?

In some other of the following essays (IX., X.,
XI., XII., XIV., XV.) I have endeavoured to
show that sober and well-founded physical and
literary criticism plays no less havoc with the
doctrine that the canonical scriptures of the New
Testament "declare incontrovertibly the actual
historical truth in all records." We are told that
the Gospels contain a true revelation of the
spiritual world—a proposition which, in one sense
of the word "spiritual," I should not think it
necessary to dispute. But, when it is taken to
signify that everything we are told about the
world of spirits in these books is infallibly true;
that we are bound to accept the demonology
which constitutes an inseparable part of their
teaching; and to profess belief in a Supernaturalism as gross as that of any primitive people—it is
at any rate permissible to ask why? Science
may be unable to define the limits of possibility,
but it cannot escape from the moral obligation
to weigh the evidence in favour of any alleged
wonderful occurrence; and I have endeavoured to
show that the evidence for the Gadarene miracle

is altogether worthless. We have simply three, partially discrepant, versions of a story, about the primitive form, the origin, and the authority for which we know absolutely nothing. But the evidence in favour of the Gadarene miracle is as good as that for any other.

Elsewhere, I have pointed out that it is utterly beside the mark to declaim against these conclusions on the ground of their asserted tendency to deprive mankind of the consolations of the Christian faith, and to destroy the foundations of morality; still less to brand them with the question-begging vituperative appellation of "infidelity." The point is not whether they are wicked; but, whether, from the point of view of scientific method, they are irrefragably true. If they are, they will be accepted in time, whether they are wicked, or not wicked. Nature, so far as we have been able to attain to any insight into her ways, recks little about consolation and makes for righteousness by very round-about paths. And, at any rate, whatever may be possible for other people, it is becoming less and less possible for the man who puts his faith in scientific methods of ascertaining truth, and is accustomed to have that faith justified by daily experience, to be consciously false to his principle in any matter. But the number of such men, driven into the use of scientific methods of inquiry and taught to trust them, by their education, their daily professional and business needs, is increasing and will continually increase. The phraseology of Supernaturalism may remain on men's lips, but in practice they are Naturalists. The magistrate who listens with devout attention to the precept "Thou shalt not suffer a witch to live" on Sunday, on Monday, dismisses, as intrinsically absurd, a charge of bewitching a cow brought against some old woman; the superintendent of a lunatic asylum who substituted exorcism for rational modes of treatment would have but a short tenure of office; even parish clerks doubt the utility of prayers for rain, so long as the wind is in the east; and an outbreak of pestilence sends men, not to the churches, but to the drains. In spite of prayers for the success of our arms and *Te Deums* for victory, our real faith is in big battalions and keeping our powder dry; in knowledge of the science of warfare; in energy, courage, and discipline. In these, as in all other practical affairs, we act on the aphorism "*Laborare est orare*"; we admit that intelligent work is the only acceptable worship; and that, whether there be a Supernature or not, our business is with Nature.

It is important to note that the principle of the scientific Naturalism of the latter half of the nineteenth century, in which the intellectual movement of the Renascence has culminated, and which was first clearly formulated by Descartes, leads not to the denial of the existence of any Supernature;[1] but simply to the denial of the validity of the evidence adduced in favour of this, or of that, extant form of Supernaturalism.

Looking at the matter from the most rigidly scientific point of view, the assumption that, amidst the myriads of worlds scattered through endless space, there can be no intelligence, as much greater than man's as his is greater than a blackbeetle's; no being endowed with powers of influencing the course of nature as much greater than his, as his is greater than a snail's, seems to me not merely baseless, but impertinent. Without stepping beyond the analogy of that which is known, it is easy to people the cosmos with entities, in ascending scale, until we reach something practically indistinguishable from omnipotence, omnipresence, and omniscience. If our intelligence can, in some matters, surely reproduce the past of thousands of years ago and anticipate the future, thousands of years hence, it is clearly within the limits of possibility that some greater intellect, even of the same order, may be able to mirror the whole past and the whole future; if the universe

---

[1] I employ the words "Supernature" and "Supernatural" in their popular senses. For myself, I am bound to say that the term "Nature" covers the totality of that which is. The world of psychical phenomena appears to me to be as much part of "Nature" as the world of physical phenomena; and I am unable to perceive any justification for cutting the Universe into two halves, one natural and one supernatural.

is penetrated by a medium of such a nature that a
magnetic needle on the earth answers to a
commotion in the sun, an omnipresent agent is
also conceivable; if our insignificant knowledge
gives us some influence over events, practical
omniscience may confer indefinably greater power.
Finally, if evidence that a thing may be, were
equivalent to proof that it is, analogy might justify
the construction of a naturalistic theology and
demonology not less wonderful than the current
supernatural; just as it might justify the peopling
of Mars, or of Jupiter, with living forms to which
terrestrial biology offers no parallel.   Until human
life is longer and the duties of the present press
less heavily, I do not think that wise men will occupy themselves with Jovian, or Martian, natural
history; and they will probably agree to a verdict
of "not proven" in respect of naturalistic theology,
taking refuge in that agnostic confession, which
appears to me to be the only position for people
who object to say that they know what they are
quite aware they do not know.   As to the interests of morality, I am disposed to think that
if mankind could be got to act up to this last
principle in every relation of life, a reformation
would be effected such as the world has not yet seen;
an approximation to the millennium, such as no
supernaturalistic religion has ever yet succeeded,
or seems likely ever to succeed, in effecting.

I have hitherto dwelt upon scientific Naturalism chiefly in its critical and destructive aspect. But the present incarnation of the spirit of the Renascence differs from its predecessor in the eighteenth century, in that it builds up, as well as pulls down.

That of which it has laid the foundation, of which it is already raising the superstructure, is the doctrine of evolution. But so many strange misconceptions are current about this doctrine—it is attacked on such false grounds by its enemies, and made to cover so much that is disputable by some of its friends, that I think it well to define as clearly as I can, what I do not and what I do understand by the doctrine.

I have nothing to say to any "Philosophy of Evolution." Attempts to construct such a philosophy may be as useful, nay, even as admirable, as was the attempt of Descartes to get at a theory of the universe by the same *a priori* road; but, in my judgment, they are as premature. Nor, for this purpose, have I to do with any theory of the "Origin of Species," much as I value that which is known as the Darwinian theory. That the doctrine of natural selection presupposes evolution is quite true; but it is not true that evolution necessarily implies natural selection. In fact, evolution might conceivably have taken place without the development of groups possessing the characters of species.

For me, the doctrine of evolution is no speculation, but a generalisation of certain facts, which
may be observed by any one who will take the
necessary trouble. These facts are those which
are classed by biologists under the heads of
Embryology and of Palæontology. Embryology
proves that every higher form of individual life
becomes what it is by a process of gradual differentiation from an extremely low form; palæontology proves, in some cases, and renders probable in
all, that the oldest types of a group are the
lowest; and that they have been followed by a
gradual succession of more and more differentiated
forms. It is simply a fact, that evolution of the
individual animal and plant is taking place, as a
natural process, in millions and millions of cases
every day; it is a fact, that the species which have
succeeded one another in the past, do, in many
cases, present just those morphological relations,
which they must possess, if they had proceeded,
one from the other, by an analogous process of
evolution.

The alternative presented, therefore, is : either
the forms of one and the same type—say, *e.g.*, that
of the Horse tribe [1]—arose successively, but independently of one another, at intervals, during
myriads of years; or, the later forms are modified

[1] The general reader will find an admirably clear and concise
statement of the evidence in this case, in Professor Flower's
recently published work *The Horse: a Study in Natural History.*

descendants of the earlier. And the latter supposition is so vastly more probable than the former,
that rational men will adopt it, unless satisfactory
evidence to the contrary can be produced. The
objection sometimes put forward, that no one yet
professes to have seen one species pass into another,
comes oddly from those who believe that mankind
are all descended from Adam. Has any one then yet
seen the production of negroes from a white stock,
or *vice versâ*? Moreover, is it absolutely necessary
to have watched every step of the progress of a
planet, to be justified in concluding that it really
does go round the sun? If so, astronomy is in a
bad way.

I do not, for a moment, presume to suggest that
some one, far better acquainted than I am with
astronomy and physics; or that a master of the
new chemistry, with its extraordinary revelations;
or that a student of the development of human
society, of language, and of religions, may not
find a sufficient foundation for the doctrine of
evolution in these several regions. On the contrary,
I rejoice to see that scientific investigation, in all
directions, is tending to the same result. And it
may well be, that it is only my long occupation
with biological matters that leads me to feel safer
among them than anywhere else. Be that as it
may, I take my stand on the facts of embryology
and of palæontology; and I hold that our present
knowledge of these facts is sufficiently thorough

and extensive to justify the assertion that all future philosophical and theological speculations will have to accommodate themselves to some such common body of established truths as the following :—

1. Plants and animals have existed on our planet for many hundred thousand, probably millions, of years. During this time, their forms, or species, have undergone a succession of changes, which eventually gave rise to the species which constitute the present living population of the earth. There is no evidence, nor any reason to suspect, that this secular process of evolution is other than a part of the ordinary course of nature; there is no more ground for imagining the occurrence of supernatural intervention, at any moment in the development of species in the past, than there is for supposing such intervention to take place, at any moment in the development of an individual animal or plant, at the present day.

2. At present, every individual animal or plant commences its existence as an organism of extremely simple anatomical structure; and it acquires all the complexity it ultimately possesses by gradual differentiation into parts of various structure and function. When a series of specific forms of the same type, extending over a long period of past time, is examined, the relation between the earlier and the later forms is analogous to that between earlier and later stages of indi-

phenomena of consciousness are manifested, it is
impossible to say.   No one doubts their presence
in his fellow-men ; and, unless any strict Cartesians
are left, no one doubts that mammals and birds are
to be reckoned creatures that have feelings analogous to our smell, taste, sight, hearing, touch,
pleasure, and pain.   For my own part, I should
be disposed to extend this analogical judgment a
good deal further.   On the other hand, if the
lowest forms of plants are to be denied consciousness, I do not see on what ground it is to be
ascribed to the lowest animals.   I find it hard to
believe that an infusory animalcule, a foraminifer,
or a fresh-water polype is capable of feeling; and,
in spite of Shakspere, I have doubts about the
great sensitiveness of the "poor beetle that we
tread upon."   The question is equally perplexing
when we turn to the stages of development of the
individual.   Granted a fowl feels; that the chick
just hatched feels; that the chick when it chirps
within the egg may possibly feel; what is to be
said of it on the fifth day, when the bird is there,
but with all its tissues nascent ?   Still more, on
the first day, when it is nothing but a flat cellular
disk ?   I certainly cannot bring myself to believe
that this disk feels.   Yet if it dose not, there must
be some time in the three weeks, between the
first day and the day of hatching, when, as a concomitant, or a consequence, of the attainment by
the brain of the chick of a certain stage of

structural evolution, consciousness makes its appearance. I have frequently expressed my incapacity to understand the nature of the relation
between consciousness and a certain anatomical
tissue, which is thus established by observation.
But the fact remains that, so far as observation and
experiment go, they teach us that the psychical
phenomena are dependent on the physical.

In like manner, if fishes, insects, scorpions, and
such animals as the pearly nautilus, possess
feeling, then undoubtedly consciousness was present in the world as far back as the Silurian
epoch. But, if the earliest animals were similar
to our rhizopods and monads, there must have
been some time, between the much earlier epoch
in which they constituted the whole animal
population and the Silurian, in which feeling
dawned, in consequence of the organism having
reached the stage of evolution on which it
depends.

5. Consciousness has various forms, which may
be manifested independently of one another.
The feelings of light and colour, of sound, of
touch, though so often associated with those of
pleasure and pain, are, by nature, as entirely
independent of them as is thinking. An animal
devoid of the feelings of pleasure and of pain,
may nevertheless exhibit all the effects of sensa-
tion and purposive action. Therefore, it would be
a justifiable hypothesis that, long after organic

evolution had attained to consciousness, pleasure
and pain were still absent. Such a world would
be without either happiness or misery; no act
could be punished and none could be rewarded;
and it could have no moral purpose.

6. Suppose, for argument's sake, that all
mammals and birds are subjects of pleasure and
pain. Then we may be certain that these forms
of consciousness were in existence at the beginning
of the Mesozoic epoch. From that time forth,
pleasure has been distributed without reference to
merit, and pain inflicted without reference to
demerit, throughout all but a mere fraction of the
higher animals. Moreover, the amount and the
severity of the pain, no less than the variety and
acuteness of the pleasure, have increased with
every advance in the scale of evolution. As
suffering came into the world, not in consequence
of a fall, but of a rise, in the scale of being, so
every further rise has brought more suffering.
As the evidence stands, it would appear that the
sort of brain which characterises the highest
mammals and which, so far as we know, is the
indispensable condition of the highest sensibility,
did not come into existence before the Tertiary
epoch. The primordial anthropoid was probably, in
this respect, on much the same footing as his pithe-
coid kin. Like them he stood upon his " natural
rights," gratified all his desires to the best of his
ability, and was as incapable of either right or

wrong doing as they. It would be as absurd as in their case, to regard his pleasures, any more than theirs, as moral rewards, and his pains, any more than theirs, as moral punishments.

7. From the remotest ages of which we have any cognizance, death has been the natural and, apparently, the necessary concomitant of life. In our hypothetical world (3), inhabited by nothing but plants, death must have very early resulted from the struggle for existence: many of the crowd must have jostled one another out of the conditions on which life depends. The occurrence of death, as far back as we have any fossil record of life, however, needs not to be proved by such arguments; for, if there had been no death there would have been no fossil remains, such as the great majority of those we met with. Not only was there death in the world, as far as the record of life takes us; but, ever since mammals and birds have been preyed upon by carnivorous animals, there has been painful death, inflicted by mechanisms specially adapted for inflicting it.

8. Those who are acquainted with the closeness of the structural relations between the human organisation and that of the mammals which come nearest to him, on the one hand; and with the palæontological history of such animals as horses and dogs, on the other; will not be disposed to question the origin of man from forms which stand in the same sort of relation to *Homo*

sapiens, as *Hipparion* does to *Equus*. I think it a conclusion, fully justified by analogy, that, sooner or later, we shall discover the remains of our less specialised primatic ancestors in the strata which have yielded the less specialised equine and canine quadrupeds. At present, fossil remains of men do not take us back further than the later part of the Quaternary epoch ; and, as was to be expected, they do not differ more from existing men, than Quaternary horses differ from existing horses. Still earlier we find traces of man, in implements, such as are used by the ruder savages at the present day. Later, the remains of the palæolithic and neolithic conditions take us gradually from the savage state to the civilisations of Egypt and of Mycenæ ; though the true chronological order of the remains actually discovered may be uncertain.

9. Much has yet to be learned, but, at present, natural knowledge affords no support to the notion that men have fallen from a higher to a lower state. On the contrary, everything points to a slow natural evolution ; which, favoured by the surrounding conditions in such localities as the valleys of the Yang-tse-kang, the Euphrates, and the Nile, reached a relatively high pitch, five or six thousand years ago ; while, in many other regions, the savage condition has persisted down to our day. In all this vast lapse of time there is not a trace of the occurrence of any general

destruction of the human race; not the smallest indication that man has been treated on any other principles than the rest of the animal world.

10. The results of the process of evolution in the case of man, and in that of his more nearly allied contemporaries, have been marvellously different. Yet it is easy to see that small primitive differences of a certain order, must, in the long run, bring about a wide divergence of the human stock from the others. It is a reasonable supposition that, in the earliest human organisms, an improved brain, a voice more capable of modulation and articulation, limbs which lent themselves better to gesture, a more perfect hand, capable among other things of imitating form in plastic or other material, were combined with the curiosity, the mimetic tendency, the strong family affection of the next lower group; and that they were accompanied by exceptional length of life and a prolonged minority. The last two peculiarities are obviously calculated to strengthen the family organisation, and to give great weight to its educative influences. The potentiality of language, as the vocal symbol of thought, lay in the faculty of modulating and articulating the voice. The potentiality of writing, as the visual symbol of thought, lay in the hand that could draw; and in the mimetic tendency, which, as we know, was gratified by drawing, as far back as the

days of Quaternary man. With speech as the record, in tradition, of the experience of more than one generation; with writing as the record of that of any number of generations; the experience of the race, tested and corrected generation after generation, could be stored up and made the starting point for fresh progress. Having these perfectly natural factors of the evolutionary process in man before us, it seems unnecessary to go further a-field in search of others.

11. That the doctrine of evolution implies a former state of innocence of mankind is quite true; but, as I have remarked, it is the innocence of the ape and of the tiger, whose acts, however they may run counter to the principles of morality, it would be absurd to blame. The lust of the one and the ferocity of the other are as much provided for in their organisation, are as clear evidences of design, as any other features that can be named.

Observation and experiment upon the phenomena of society soon taught men that, in order to obtain the advantages of social existence, certain rules must be observed. Morality commenced with society. Society is possible only upon the condition that the members of it shall surrender more or less of their individual freedom of action. In primitive societies, individual selfishness is a centrifugal force of such intensity that it is

constantly bringing the social organisation to the
verge of destruction.   Hence the prominence of
the positive rules of obedience to the elders; of
standing by the family or the tribe in all emergen-
cies; of fulfilling the religious rites, non-observ-
ance of which is conceived to damage it with the
supernatural powers, belief in whose existence is
one of the earliest products of human thought;
and of the negative rules, which restrain each
from meddling with the life or property of
another.

12. The highest conceivable form of human
society is that in which the desire to do what is
best for the whole, dominates and limits the
action of every member of that society.   The
more complex the social organisation the greater
the number of acts from which each man must
abstain, if he desires to do that which is best for
all.   Thus the progressive evolution of society
means increasing restriction of individual freedom
in certain directions.

With the advance of civilisation, and the
growth of cities and of nations by the coalescence
of families and of tribes, the rules which constitute the common foundation of morality and of
law became more numerous and complicated, and
the temptations to break or evade many of them
stronger.   In the absence of a clear apprehension of the natural sanctions of these rules, a
supernatural sanction was assumed; and imagina-

tion supplied the motives which reason was
supposed to be incompetent to furnish. Religion,
at first independent of morality, gradually took
morality under its protection; and the supernaturalists have ever since tried to persuade
mankind that the existence of ethics is bound up
with that of supernaturalism.

I am not of that opinion, But, whether it is
correct or otherwise, it is very clear to me that,
as Beelzebub is not to be cast out by the aid of
Beelzebub, so morality is not to be established
by immorality. It is, we are told, the special
peculiarity of the devil that he was a liar from
the beginning. If we set out in life with pre-
tending to know that which we do not know; with
professing to accept for proof evidence which we
are well aware is inadequate; with wilfully
shutting our eyes and our ears to facts which
militate against this or that comfortable hypothesis; we are assuredly doing our best to deserve
the same character.

I have not the presumption to imagine that, in
spite of all my efforts, errors may not have crept
into these propositions. But I am tolerably
confident that time will prove them to be
substantially correct. And if they are so, I
confess I do not see how any extant supernaturalistic system can also claim exactness. That they
are irreconcilable with the biblical cosmogony,

anthropology, and theodicy is obvious; but they
are no less inconsistent with the sentimental
Deism of the "Vicaire Savoyard" and his
numerous modern progeny. It is as impossible,
to my mind, to suppose that the evolutionary
process was set going with full foreknowledge of
the result and yet with what we should understand by a purely benevolent intention, as it is
to imagine that the intention was purely malevo-
lent. And the prevalence of dualistic theories
from the earliest times to the present day—
whether in the shape of the doctrine of the
inherently evil nature of matter; of an Ahriman;
of a hard and cruel Demiurge; of a diabolical
"prince of this world," show how widely this
difficulty has been felt.

Many seem to think that, when it is admitted
that the ancient literature, contained in our
Bibles, has no more claim to infallibility than any
other ancient literature; when it is proved that
the Israelites and their Christian successors
accepted a great many supernaturalistic theories
and legends which have no better foundation than
those of heathenism, nothing remains to be done but
to throw the Bible aside as so much waste paper.

I have always opposed this opinion. It appears
to me that if there is anybody more objectionable
than the orthodox Bibliolater it is the heterodox
Philistine, who can discover in a literature which,
in some respects, has no superior, nothing but

a subject for scoffing and an occasion for the display of his conceited ignorance of the debt he owes to former generations.

Twenty-two years ago I pleaded for the use of the Bible as an instrument of popular education, and I venture to repeat what I then said:

"Consider the great historical fact that, for three centuries, this book has been woven into the life of all that is best and noblest in English history; that it has become the national Epic of Britain and is as familiar to gentle and simple, from John o' Groat's House to Land's End, as Dante and Tasso once were to the Italians; that it is written in the noblest and purest English and abounds in exquisite beauties of mere literary form; and, finally, that it forbids the veriest hind, who never left his village, to be ignorant of the existence of other countries and other civilisations and of a great past, stretching back to the furthest limits of the oldest nations in the world. By the study of what other book could children be so much humanised and made to feel that each figure in that vast historical procession fills, like themselves, but a momentary space in the interval between the Eternities; and earns the blessings or the curses of all time, according to its effort to do good and hate evil, even as they also are earning their payment for their work?"[1]

[1] "The School Boards: What they Can do and what they May do," 1870. *Critiques and Addresses*, p. 51.

At the same time, I laid stress upon the necessity of placing such instruction in lay hands; in the hope and belief, that it would thus gradually accommodate itself to the coming changes of opinion; that the theology and the legend would drop more and more out of sight, while the perennially interesting historical, literary, and ethical contents would come more and more into view.

I may add yet another claim of the Bible to the respect and the attention of a democratic age. Throughout the history of the western world, the Scriptures, Jewish and Christian, have been the great instigators of revolt against the worst forms of clerical and political despotism. The Bible has been the *Magna Charta* of the poor and of the oppressed; down to modern times, no State has had a constitution in which the interests of the people are so largely taken into account, in which the duties, so much more than the privileges, of rulers are insisted upon, as that drawn up for Israel in Deuteronomy and in Leviticus; nowhere is the fundamental truth that the welfare of the State, in the long run, depends on the uprightness of the citizen so strongly laid down. Assuredly, the Bible talks no trash about the rights of man; but it insists on the equality of duties, on the liberty to bring about that righteousness which is somewhat different from struggling for "rights"; on the fraternity of taking thought for one's neighbour as for one's self.

the solace of fame, if not by rewards of a less elevated character.

So, following the advice of Francis Bacon, we refuse *inter mortuos quærere vivum;* we leave the past to bury its dead, and ignore our intellectual ancestry. Nor are we content with that. We follow the evil example set us, not only by Bacon but by almost all the men of the Renaissance, in pouring scorn upon the work of our immediate spiritual forefathers, the schoolmen of the Middle Ages. It is accepted as a truth which is indisputable, that, for seven or eight centuries, a long succession of able men—some of them of transcendent acuteness and encyclopædic knowledge—devoted laborious lives to the grave discussion of mere frivolities and the arduous pursuit of intellectual will-o'-the-wisps. To say nothing of a little modesty, a little impartial pondering over personal experience might suggest a doubt as to the adequacy of this short and easy method of dealing with a large chapter of the history of the human mind. Even an acquaintance with popular literature which had extended so far as to include that part of the contributions of Sam Slick which contains his weighty aphorism that "there is a great deal of human nature in all mankind," might raise a doubt whether, after all, the men of that epoch, who, take them all round, were endowed with wisdom and folly in much the same proportion as ourselves, were likely to display nothing better than the qualities of energetic idiots, when they devoted their faculties to the elucidation of problems which were to them, and indeed are to us, the most serious which life has to offer. Speaking for myself, the longer I live the more I am disposed to think that there is much less either of pure folly, or of pure wickedness, in the world than is commonly supposed. It may be doubted if any sane man ever said to himself, "Evil, be thou my good," and have never yet had the good fortune to meet with a perfect fool. When I have brought to the inquiry the patience and long-suffering which become a scientific investigator, the most promising specimens have turned out to have a good deal to say for themselves from their own point of view. And, sometimes, calm reflection has taught the humiliating lesson, that their point of view was not so different from my own as I had fondly imagined. Comprehension is more than half-way to sympathy, here as elsewhere.

If we turn our attention to scholastic philosophy in the frame of mind suggested by these prefatory remarks, it assumes a very different character from that which it bears in general estimation. No doubt it is surrounded by a dense thicket of thorny logomachies and obscured by the dust-clouds of a barbarous and perplexing terminology. But suppose that, undeterred by much grime and

by many scratches, the explorer has toiled through
this jungle, he comes to an open country which is
amazingly like his dear native land. The hills
which he has to climb, the ravines he has to
avoid, look very much the same; there is the
same infinite space above, and the same abyss of
the unknown below; the means of travelling are
the same, and the goal is the same.

That goal for the schoolmen, as for us, is the
settlement of the question how far the universe is
the manifestation of a rational order; in other
words, how far logical deduction from indisputable premisses will account for that which has
happened and does happen. That was the object
of scholasticism, and, so far as I am aware, the
object of modern science may be expressed in
the same terms. In pursuit of this end, modern
science takes into account all the phenomena of
the universe which are brought to our knowledge
by observation or by experiment. It admits that
there are two worlds to be considered, the one
physical and the other psychical; and that though
there is a most intimate relation and interconnec-
tion between the two, the bridge from one to the
other has yet to be found; that their phenomena
run, not in one series, but along two parallel lines.

To the schoolmen the duality of the universe
appeared under a different aspect. How this
came about will not be intelligible unless we
clearly apprehend the fact that they did really

believe in dogmatic Christianity as it was formulated by the Roman Church. They did not give
a mere dull assent to anything the Church told
them on Sundays, and ignore her teachings for
the rest of the week; but they lived and moved
and had their being in that supersensible theological world which was created, or rather grew
up, during the first four centuries of our reckoning,
and which occupied their thoughts far more than
the sensible world in which their earthly lot was
cast.

For the most part, we learn history from the
colourless compendiums or partisan briefs of mere
scholars, who have too little acquaintance with
practical life, and too little insight into speculative problems, to understand that about which
they write. In historical science, as in all
sciences which have to do with concrete phenomena, laboratory practice is indispensable; and
the laboratory practice of historical science is
afforded, on the one hand, by active social and
political life, and, on the other, by the study of
those tendencies and operations of the mind which
embody themselves in philosophical and theological systems. Thucydides and Tacitus, and, to come
nearer our own time, Hume and Grote, were men
of affairs, and had acquired, by direct contact with
social and political history in the making, the
secret of understanding how such history is made.
Our notions of the intellectual history of the

middle ages are, unfortunately, too often derived from writers who have never seriously grappled with philosophical and theological problems: and hence that strange myth of a millennium of moonshine to which I have adverted.

However, no very profound study of the works of contemporary writers who, without devoting themselves specially to theology or philosophy, were learned and enlightened—such men, for example, as Eginhard or Dante—is necessary to convince one's self, that, for them, the world of the theologian was an ever-present and awful reality. From the centre of that world, the Divine Trinity, surrounded by a hierarchy of angels and saints, contemplated and governed the insignificant sensible world in which the inferior spirits of men, burdened with the debasement of their material embodiment and continually solicited to their perdition by a no less numerous and almost as powerful hierarchy of devils, were constantly struggling on the edge of the pit of everlasting damnation.[1]

[1] There is no exaggeration in this brief and summary view of the Catholic cosmos. But it would be unfair to leave it to be supposed that the Reformation made any essential alteration, except perhaps for the worse, in that cosmology which called itself "Christian." The protagonist of the Reformation, from whom the whole of the Evangelical sects are lineally descended, states the case with that plainness of speech, not to say brutality, which characterised him. Luther says that man is a beast of burden who only moves as his rider orders; sometimes God rides him, and sometimes Satan. "Sic voluntas humana in medio posita est, ceu jumentum; si insederit Deus, vult et

The men of the middle ages believed that through the Scriptures, the traditions of the Fathers, and the authority of the Church, they were in possession of far more, and more trustworthy, information with respect to the nature and order of things in the theological world than they had in regard to the nature and order of things in the sensible world. And, if the two sources of information came into conflict, so much the worse for the sensible world, which, after all, was more or less under the dominion of Satan. Let us suppose that a telescope powerful enough to show us what is going on in the nebula of the sword of Orion, should reveal a world in which stones fell upwards, parallel lines met, and the fourth dimension of space was quite obvious. Men of science would have only two alternatives before them. Either the terrestrial and the nebular facts must be brought into harmony by such feats of subtle sophistry as the human mind is always

vadit, quo vult Deus. . . . Si insederit Satan, vult et vadit, quo vult Satan; nec est in ejus arbitrio ad utrum sessorem currere, aut eum quærere, sed ipsi sessores certant ob ipsum obtinendum et possidendum " (*De Servo Arbitrio*, M. Lutheri Opera, ed. 1546, t. ii. p. 468). One may hear substantially the same doctrine preached in the parks and at street-corners by zealous volunteer missionaries of Evangelicism, any Sunday, in modern London. Why these doctrines, which are conspicuous by their absence in the four Gospels, should arrogate to themselves the title of Evangelical, in contradistinction to Catholic, Christianity, may well perplex the impartial inquirer, who, if he were obliged to choose between the two, might naturally prefer that which leaves the poor beast of burden a little freedom of choice.

capable of performing when driven into a corner;
or science must throw down its arms in despair,
and commit suicide, either by the admission that
the universe is, after all, irrational, inasmuch as
that which is truth in one corner of it is absurdity
in another, or by a declaration of incompetency.

In the middle ages, the labours of those great
men who endeavoured to reconcile the system of
thought which started from the data of pure
reason, with that which started from the data of
Roman theology, produced the system of thought
which is known as scholastic philosophy; the
alternative of surrender and suicide is exemplified
by Avicenna and his followers when they declared
that that which is true in theology may be false
in philosophy, and *vice versâ;* and by Sanchez
in his famous defence of the thesis "*Quod nil
scitur.*"

To those who deny the validity of one of the
primary assumptions of the disputants—who
decline, on the ground of the utter insufficiency of
the evidence, to put faith in the reality of that
other world, the geography and the inhabitants of
which are so confidently described in the so-called[1]
Christianity of Catholicism—the long and bitter
contest, which engaged the best intellects for so

---

[1] I say "so-called" not by way of offence, but as a protest
against the monstrous assumption that Catholic Christianity is
explicitly or implicitly contained in any trustworthy record of
the teaching of Jesus of Nazareth.

many centuries, may seem a terrible illustration
of the wasteful way in which the struggle for existence is carried on in the world of thought, no
less than in that of matter. But there is a more
cheerful mode of looking at the history of scholasticism. It ground and sharpened the dialectic
implements of our race as perhaps nothing but
discussions, in the result of which men thought
their eternal, no less than their temporal, interests·
were at stake, could have done. When a logical
blunder may ensure combustion, not only in the
next world but in this, the construction of syllo-
gisms acquires a peculiar interest. Moreover, the
schools kept the thinking faculty alive and active,
when the disturbed state of civil life, the mephitic
atmosphere engendered by the dominant ecclesiasticism, and the almost total neglect of natural
knowledge, might well have stifled it. And,
finally, it should be remembered that scholasticism
really did thresh out pretty effectually certain
problems which have presented themselves to
mankind ever since they began to think, and
which, I suppose, will present themselves so long
as they continue to think. Consider, for example,
the controversy of the Realists and the Nominal-
ists, which was carried on with varying fortunes,
and under various names, from the time of Scotus
Erigena to the end of the scholastic period. Has
it now a merely antiquarian interest? Has
Nominalism, in any of its modifications, so com-

quite in accordance with science by many excellent, instructed, and intelligent people.

The preacher further contended that it was yet more difficult to realise that our earthly home would become the scene of a vast physical catastrophe. Imagination recoils from the idea that the course of nature—the phrase helps to disguise the truth —so unvarying and regular, the ordered sequence of movement and life, should suddenly cease. Imagination looks more reasonable when it assumes the air of scientific reason. Physical law, it says, will prevent the occurrence of catastrophes only anticipated by an apostle in an unscientific age. Might not there, however, be a suspension of a lower law by the intervention of a higher? Thus every time we lifted our arms we defied the laws of gravitation, and in railways and steamboats powerful laws were held in check by others. The flood and the destruction of Sodom and Gomorrah were brought about by the operations of existing laws, and may it not be that in His illimitable universe there are more important laws than those which surround our puny life—moral and not merely physical forces? Is it inconceivable that the day will come when these royal and ultimate laws shall wreck the natural order of things which seems so stable and so fair? Earthquakes were not things of remote antiquity, as an island off Italy, the Eastern Archipelago, Greece, and Chicago bore witness. . . . In presence of a great earthquake men feel how powerless they are, and their very knowledge adds to their weakness. The end of human probation, the final dissolution of organised society, and the destruction of man's home on the surface of the globe, were none of them violently contrary to our present experience, but only the extension of present facts. The presentiment of death was common ; there were felt to be many things which threatened the existence of society ; and as our globe was a ball of fire, at any moment the pent-up forces which surge and boil beneath our feet might be poured out (" Pall Mall Gazette," December 6, 1886).

The preacher appears to entertain the notion

that the occurrence of a "catastrophe" [1] involves a breach of the present order of nature—that it is an event incompatible with the physical laws which at present obtain. He seems to be of opinion that "scientific reason" lends its authority to the imaginative supposition that physical law will prevent the occurrence of the "catastrophes" anticipated by an unscientific apostle.

Scientific reason, like Homer, sometimes nods; but I am not aware that it has ever dreamed dreams of this sort. The fundamental axiom of scientific thought is that there is not, never has been, and never will be, any disorder in nature. The admission of the occurrence of any event which was not the logical consequence of the immediately antecedent events, according to these definite, ascertained, or unascertained rules which we call the "laws of nature," would be an act of self-destruction on the part of science.

"Catastrophe" is a relative conception. For ourselves it means an event which brings about very terrible consequences to man, or impresses his mind by its magnitude relatively to him. But events which are quite in the natural order of things to us, may be frightful catastrophes to other sentient beings. Surely no interruption of the

---

[1] At any rate a catastrophe greater than the flood, which, as I observe with interest, is as calmly assumed by the preacher to be an historical event as if science had never had a word to say on that subject !

order of nature is involved if, in the course of
descending through an Alpine pine-wood, I jump
upon an anthill and in a moment wreck a whole
city and destroy a hundred thousand of its inhabitants. To the ants the catastrophe is worse than
the earthquake of Lisbon. To me it is the natural
and necessary consequence of the laws of matter
in motion. A redistribution of energy has taken
place, which is perfectly in accordance with
natural order, however unpleasant its effects may
be to the ants.

Imagination, inspired by scientific reason,
and not merely assuming the airs thereof, as it
unfortunately too often does in the pulpit, so far
from having any right to repudiate catastrophes
and deny the possibility of the cessation of motion
and life, easily finds justification for the exactly
contrary course. Kant in his famous " Theory of the
Heavens " declares the end of the world and its
reduction to a formless condition to be a necessary
consequence of the causes to which it owes its
origin and continuance.   And, as to catastrophes of prodigious magnitude and frequent occurrence, they were the favourite *asylum ignorantiæ*
of geologists, not a quarter of a century ago.   If
modern geology is becoming more and more
disinclined to call in catastrophes to its aid, it is
not because of any *a priori* difficulty in reconciling
the occurrence of such events with the universality
of order, but because the *a posteriori* evidence of

the occurrence of events of this character in past times has more or less completely broken down.

It is, to say the least, highly probable that this earth is a mass of extremely hot matter, invested by a cooled crust, through which the hot interior still continues to cool, though with extreme slowness. It is no less probable that the faults and dislocations, the foldings and fractures, everywhere visible in the stratified crust, its large and slow movements through miles of elevation and depression, and its small and rapid movements which give rise to the innumerable perceived and unperceived earthquakes which are constantly occurring, are due to the shrinkage of the crust on its cooling and contracting nucleus.

Without going beyond the range of fair scientific analogy, conditions are easily conceivable which should render the loss of heat far more rapid than it is at present; and such an occurrence would be just as much in accordance with ascertained laws of nature, as the more rapid cooling of a red-hot bar, when it is thrust into cold water, than when it remains in the air. But much more rapid cooling might entail a shifting and re-arrangement of the parts of the crust of the earth on a scale of unprecedented magnitude, and bring about "catastrophes" to which the earthquake of Lisbon is but a trifle. It is conceivable that man and his works and all the higher forms of animal life should be utterly destroyed; that mountain regions should be converted into ocean depths and the floor of oceans raised into mountains; and the earth become a scene of horror which even the lurid fancy of the writer of the Apocalypse would fail to portray. And yet, to the eye of science, there would be no more disorder here than in the sabbatical peace of a summer sea. Not a link in the chain of natural causes and effects would be broken, nowhere would there be the slightest indication of the "suspension of a lower law by a higher." If a sober scientific thinker is inclined to put little faith in the wild vaticinations of universal ruin which, in a less saintly person than the seer of Patmos, might seem to be dictated by the fury of a revengeful fanatic, rather than by the spirit of the teacher who bid men love their enemies, it is not on the ground that they contradict scientific principles; but because the evidence of their scientific value does not fulfil the conditions on which weight is attached to evidence. The imagination which supposes that it does, simply does not "assume the air of scientific reason."

I repeat that, if imagination is used within the limits laid down by science, disorder is unimaginable. If a being endowed with perfect intellectual and æsthetic faculties, but devoid of the capacity for suffering pain, either physical or moral, were to devote his utmost powers to the investigation of nature, the universe would seem to him to be a sort of kaleidoscope, in which, at every successive moment of time, a new arrangement of parts of exquisite beauty and symmetry would present itself; and each of them would show itself to be the logical consequence of the preceding arrangement, under the conditions which we call the laws of nature. Such a spectator might well be filled with that *Amor intellectualis Dei*, the beatific vision of the *vita contemplativa*, which some of the greatest thinkers of all ages, Aristotle, Aquinas, Spinoza, have regarded as the only conceivable eternal felicity; and the vision of illimitable suffering, as if sensitive beings were unregarded animalcules which had got between the bits of glass of the kaleidoscope, which mars the prospect to us poor mortals, in no wise alters the fact that order is lord of all, and disorder only a name for that part of the order which gives us pain.

The other fallacious employment of the names of scientific conceptions which pervades the preacher's utterance, brings me back to the proper topic of the present essay. It is the use of the word "law" as if it denoted a thing—as if a "law of nature," as science understands it, were a being endowed with certain powers, in virtue of which the phenomena expressed by that law are brought about. The preacher asks, "Might not there be a suspension of a lower law by the intervention of a higher?" He tells us that every time we lift our arms we defy the law of gravitation. He asks whether some day certain "royal and ultimate laws" may not come and "wreck" those laws which are at present, it would appear, acting as nature's police. It is evident, from these expressions, that "laws," in the mind of the preacher, are entities having an objective existence in a graduated hierarchy. And it would appear that the "royal laws" are by no means to be regarded as constitutional royalties : at any moment, they may, like Eastern despots, descend in wrath among the middle-class and plebeian laws, which have hitherto done the drudgery of the world's work, and, to use phraseology not unknown in our seats of learning—"make hay" of their belongings. Or perhaps a still more familiar analogy has suggested this singular theory; and it is thought that high laws may "suspend" low laws, as a bishop may suspend a curate.

Far be it from me to controvert these views, if any one likes to hold them. All I wish to remark is that such a conception of the nature of "laws" has nothing to do with modern science. It is scholastic realism—realism as intense and unmitigated as that of Scotus Erigena a thousand years ago. The essence of such realism is that it maintains the objective existence of universals, or, as we call them nowadays, general propositions. It affirms, for example, that "man" is a real thing, apart from individual men, having its existence, not in the sensible, but in the intelligible

world, and clothing itself with the accidents of
sense to make the Jack and Tom and Harry
whom we know.   Strange as such a notion may
appear to modern scientific thought, it really
pervades ordinary language.   There are few
people who would, at once, hesitate to admit that
colour, for example, exists apart from the mind
which conceives the idea of colour.   They hold it
to be something which resides in the coloured
object ; and so far they are as much Realists as if
they had sat at Plato's feet.   Reflection on the
facts of the case must, I imagine, convince every
one that " colour " is—not a mere name, which
was the extreme Nominalist position—but a name
for that group of states of feeling which we call
blue, red, yellow, and so on, and which we believe
to be caused by luminiferous vibrations which
have not the slightest resemblance to colour ;
while these again are set afoot by states of the
body to which we ascribe colour, but which are
equally devoid of likeness to colour.

In the same way, a law of nature, in the scientific sense, is the product of a mental operation
upon the facts of nature which come under our
observation, and has no more existence outside
the mind than colour has.   The law of gravitation
is a statement of the manner in which experience
shows that bodies, which are free to move, do, in
fact, move towards one another.   But the other
facts of observation, that bodies are not always

does happen, and our anticipation of that which will happen, is an interesting psychological fact; and would be unintelligible if the tendency of the human mind towards realism were less strong.

Even at the present day, and in the writings of men who would at once repudiate scholastic realism in any form, " law " is often inadvertently employed in the sense of cause, just as, in common life, a man will say that he is compelled by the law to do so and so, when, in point of fact, all he means is that the law orders him to do it, and tells him what will happen if he does not do it. We commonly hear of bodies falling to the ground by reason of the law of gravitation, whereas that law is simply the record of the fact that, according to all experience, they have so fallen (when free to move), and of the grounds of a reasonable expectation that they will so fall. If it should be worth anybody's while to seek for examples of such misuse of language on my own part, I am not at all sure he might not succeed, though I have usually been on my guard against such looseness of expression. If I am guilty, I do penance beforehand, and only hope that I may thereby deter others from committing the like fault. And I venture on this personal observation by way of showing that I have no wish to bear hardly on the preacher for falling into an error for which he might find good precedents. But it is one of those errors which, in the case of a person engaged in scientific pursuits, do little harm, because it is corrected as soon as its consequences become obvious; while those who know physical science only by name are, as has been seen, easily led to build a mighty fabric of unrealities on this fundamental fallacy. In fact, the habitual use of the word "law," in the sense of an active thing, is almost a mark of pseudo-science; it characterises the writings of those who have appropriated the forms of science without knowing anything of its substance.

There are two classes of these people : those who are ready to believe in any miracle so long as it is guaranteed by ecclesiastical authority; and those who are ready to believe in any miracle so long as it has some different guarantee. The believers in what are ordinarily called miracles— those who accept the miraculous narratives which they are taught to think are essential elements of religious doctrine—are in the one category; the spirit-rappers, table-turners, and all the other devotees of the occult sciences of our day are in the other: and, if they disagree in most things they agree in this, namely, that they ascribe to science a dictum that is not scientific; and that they endeavour to upset the dictum thus foisted on science by a realistic argument which is equally unscientific.

It is asserted, for example, that, on a particular occasion, water was turned into wine; and, on the other hand, it is asserted that a man or a woman "levitated" to the ceiling, floated about there, and finally sailed out by the window. And it is assumed that the pardonable scepticism, with which most scientific men receive these statements, is due to the fact that they feel themselves justified in denying the possibility of any such metamorphosis of water, or of any such levitation, because such events are contrary to the laws of nature. So the question of the preacher is triumphantly put: How do you know that there are not "higher" laws of nature than your chemical and physical laws, and that these higher laws may not intervene and "wreck" the latter?

The plain answer to this question is, Why should anybody be called upon to say how he knows that which he does not know? You are assuming that laws are agents—efficient causes of that which happens—and that one law can interfere with another. To us, that assumption is as nonsensical as if you were to talk of a proposition of Euclid being the cause of the diagram which illustrates it, or of the integral calculus interfering with the rule of three. Your question really implies that we pretend to complete knowledge not only of all past and present phenomena, but of all that are possible in the future, and we leave all that sort of thing to the adepts of esoteric Buddhism. Our pretensions are infinitely more modest. We have succeeded in finding out the rules of action of a little bit of the universe; we call these rules " laws of nature," not because anybody knows whether they bind nature or not, but because we find it is obligatory on us to take them into account, both as actors under nature, and as interpreters of nature. We have any quantity of genuine miracles of our own, and if you will furnish us with as good evidence of your miracles as we have of ours, we shall be quite happy to accept them and to amend our expression of the laws of nature in accordance with the new facts.

As to the particular cases adduced, we are so perfectly fair-minded as to be willing to help your case as far as we can. You are quite mistaken in supposing that anybody who is acquainted with the possibilities of physical science will undertake categorically to deny that water may be turned into wine. Many very competent judges are already inclined to think that the bodies, which we have hitherto called elementary, are really composite arrangements of the particles of a uniform primitive matter. Supposing that view to be correct, there would be no more theoretical diffi- culty about turning water into alcohol, ethereal and colouring matters, than there is, at this pres- ent moment, any practical difficulty in working other such miracles; as when we turn sugar into alcohol, carbonic acid, glycerine, and succinic acid; or transmute gas-refuse into perfumes rarer than musk and dyes richer than Tyrian purple. If the so-called " elements," oxygen and hydrogen, which compose water, are aggregates of the same ultimate particles, or physical units, as those which enter into the structure of the so-called element " carbon," it is obvious that alcohol and other substances, composed of carbon, hydrogen, and oxygen, may be produced by a rearrangement of some of the units of oxygen and hydrogen into the " element " carbon, and their synthesis with the rest of the oxygen and hydrogen.

Theoretically, therefore, we can have no sort of objection to your miracle. And our reply to the levitators is just the same. Why should not your friend " levitate " ? Fish are said to rise and sink in the water by altering the volume of an internal air-receptacle; and there may be many ways science, as yet, knows nothing of, by which we, who live at the bottom of an ocean of air, may do the same thing. Dialectic gas and wind appear to be by no means wanting among you, and why should not long practice in pneumatic philosophy have resulted in the internal generation of something a thousand times rarer than hydrogen, by which, in accordance with the most ordinary natural laws, you would not only rise to the ceiling and float there in quasi-angelic posture, but perhaps, as one of your feminine adepts is said to have done, flit swifter than train or telegram to " still-vexed Bermoothes," and twit Ariel, if he happens to be

there, for a sluggard ?   We have not the presumption to deny the possibility of anything you affirm ;
only, as our brethren are particular about evidence,
do give us as much to go upon as may save us from
being roared down by their inextinguishable
laughter.

Enough of the realism which clings about " laws."
There are plenty of other exemplifications of its
vitality in modern science, but I will cite only one
of them.

This is the conception of " vital force " which
comes straight from the philosophy of Aristotle.
It is a fundamental proposition of that philosophy
that a natural object is composed of two constitu-
ents—the one its matter, conceived as inert or
even, to a certain extent, opposed to orderly and
purposive motion ; the other its form, conceived as
a quasi-spiritual something, containing or con-
ditioning the actual activities of the body and the
potentiality of its possible activities.

I am disposed to think that the prominence of
this conception in Aristotle's theory of things
arose from the circumstance that he was, to begin
with and throughout his life, devoted to biological
studies.   In fact it is a notion which must force
itself upon the mind of any one who studies
biological phenomena, without reference to general
physics, as they now stand.   Everybody who
observes the obvious phenomena of the development of a seed into a tree, or of an egg into an

animal, will note that a relatively formless mass of matter gradually grows, takes a definite shape and structure, and, finally, begins to perform actions which contribute towards a certain end, namely, the maintenance of the individual in the first place, and of the species in the second. Starting from the axiom that every event has a cause, we have here the *causa finalis* manifested in the last set of phenomena, the *causa materialis* and *formalis* in the first, while the existence of a *causa efficiens* within the seed or egg and its product, is a corollary from the phenomena of growth and metamorphosis, which proceed in unbroken succession and make up the life of the animal or plant.

Thus, at starting, the egg or seed is matter having a " form " like all other material bodies. But this form has the peculiarity, in contradistinction to lower substantial " forms," that it is a power which constantly works towards an end by means of living organisation.

So far as I know, Leibnitz is the only philosopher (at the same time a man of science, in the modern sense, of the first rank) who has noted that the modern conception of Force, as a sort of atmosphere enveloping the particles of bodies, and having potential or actual activity, is simply a new name for the Aristotelian Form.[1]   In modern biology, up till within quite recent times, the Aristotelian con-

[1] " Les formes des anciens ou Entéléchies ne sont autre chose que les forces " (Leibnitz, *Lettre au Père Bouvet*, 1697).

ception held undisputed sway; living matter was
endowed with "vital force," and that accounted for
everything. Whosoever was not satisfied with
that explanation was treated to that very "plam
argument"—"confound you eternally"—where-
with Lord Peter overcomes the doubts of his
brothers in the "Tale of a Tub" "Materialist" was
the mildest term applied to him—fortunate if he
escaped pelting with "infidel" and "atheist."
There may be scientific Rip Van Winkles about,
who still hold by vital force; but among those
biologists who have not been asleep for the last
quarter of a century "vital force" no longer
figures in the vocabulary of science. It is a patent
survival of realism; the generalisation from experience that all living bodies exhibit certain
activities of a definite character is made the basis
of the notion that every living body contains an
entity, "vital force," which is assumed to be the
cause of those activities.

It is remarkable, in looking back, to notice to
what an extent this and other survivals of
scholastic realism arrested or, at any rate, impeded
the application of sound scientific principles to
the investigation of biological phenomena. When
I was beginning to think about these matters, the
scientific world was occasionally agitated by
discussions respecting the nature of the "species"
and "genera" of Naturalists, of a different order
from the disputes of a later time. I think most

were agreed that a "species" was something which existed objectively, somehow or other, and had been created by a Divine fiat. As to the objective reality of genera, there was a good deal of difference of opinion. On the other hand, there were a few who could see no objective reality in anything but individuals, and looked upon both species and genera as hypostatised universals. As for myself, I seem to have unconsciously emulated William of Occam, inasmuch as almost the first public discourse I ever ventured upon, dealt with "Animal Individuality," and its tendency was to fight the Nominalist battle even in that quarter.

Realism appeared in still stranger forms at the time to which I refer. The community of plan which is observable in each great group of animals was hypostatised into a Platonic idea with the appropriate name of "archetype," and we were told, as a disciple of Philo-Judæus might have told us, that this realistic figment was "the archetypal light" by which Nature has been guided amidst the "wreck of worlds." So, again, another naturalist, who had no less earned a well-deserved reputation by his contributions to positive knowledge, put forward a theory of the production of living things which, as nearly as the increase of knowledge allowed, was a reproduction of the doctrine inculcated by the Jewish Cabbala.

Annexing the archetype notion, and carrying it to its full logical consequence, the author of this

theory conceived that the species of animals and
plants were so many incarnations of the thoughts
of God—material representations of Divine
ideas—during the particular period of the world's
history at which they existed. But, under the
influence of the embryological and palæontological
discoveries of modern times, which had already
lent some scientific support to the revived ancient
theories of cosmical evolution or emanation, the
ingenious author of this speculation, while denying
and repudiating the ordinary theory of evolution
by successive modification of individuals, maintained and endeavoured to prove the occurrence
of a progressive modification in the divine ideas
of successive epochs.

On the foundation of a supposed elevation of
organisation in the whole living population of any
epoch, as compared with that of its predecessor,
and a supposed complete difference in species
between the populations of any two epochs
(neither of which suppositions has stood the test
of further inquiry), the author of this speculation
based his conclusion that the Creator had, so to
speak, improved upon his thoughts as time went
on ; and that, as each such amended scheme of
creation came up, the embodiment of the earlier
divine thoughts was swept away by a universal
catastrophe, and an incarnation of the improved
ideas took its place. Only after the last such
" wreck " thus brought about, did the embodiment

of a divine thought, in the shape of the first man, make its appearance as the *ne plus ultra* of the cosmogonical process.

I imagine that Louis Agassiz, the genial back-woodsman of the science of my young days, who did more to open out new tracks in the scientific forest than most men, would have been much surprised to learn that he was preaching the doctrine of the Cabbala, pure and simple. According to this modification of Neoplatonism by contact with Hebrew speculation, the divine essence is unknowable—without form or attribute; but the interval between it and the world of sense is filled by intelligible entities, which are nothing but the familiar hypostatised abstractions of the realists. These have emanated, like immense waves of light, from the divine centre, and, as ten consecutive zones of Sephiroth, form the universe. The farther away from the centre, the more the primitive light wanes, until the periphery ends in those mere negations, darkness and evil, which are the essence of matter. On this, the divine agency transmitted through the Sephiroth operates after the fashion of the Aristotelian forms, and, at first, produces the lowest of a series of worlds. After a certain duration the primitive world is demolished and its fragments used up in making a better; and this process is repeated, until at length a final world, with man for its crown and finish, makes its appearance.

It is needless to trace the process of retrogressive metamorphosis by which, through the agency of the Messiah, the steps of the process of evolution here sketched are retraced. Sufficient has been said to prove that the extremist realism current in the philosophy of the thirteenth century can be fully matched by the speculations of our own time.

III

SCIENCE AND PSEUDO-SCIENCE

[1887]

In the opening sentences of a contribution to the
last number of this Review, [1] the Duke of Argyll
has favoured me with a lecture on the proprieties
of controversy, to which I should be disposed to
listen with more docility if his Grace's precepts
appeared to me to be based upon rational principles,
or if his example were more exemplary.

With respect to the latter point, the Duke has
thought fit to entitle his article "Professor Huxley
on Canon Liddon," and thus forces into prominence
an element of personality, which those who read
the paper which is the object of the Duke's
animadversions will observe I have endeavoured,
most carefully, to avoid. My criticisms dealt with
a report of a sermon, published in a newspaper,
and thereby addressed to all the world. Whether
that sermon was preached by A or B was not a

---

[1] *Nineteenth Century,* March 1887.

matter of the smallest consequence; and I went out of my way to absolve the learned divine to whom the discourse was attributed, from the responsibility for statements which, for anything I knew to the contrary, might contain imperfect, or inaccurate, representations of his views. The assertion that I had the wish, or was beset, by any "temptation to attack" Canon Liddon is simply contrary to fact.

But suppose that if, instead of sedulously avoiding even the appearance of such attack, I had thought fit to take a different course ; suppose that, after satisfying myself that the eminent clergyman whose name is paraded by the Duke of Argyll had really uttered the words attributed to him from the pulpit of St. Paul's, what right would any one have to find fault with my action on grounds either of justice, expediency, or good taste ?

Establishment has its duties as well as its rights. The clergy of a State Church enjoy many advantages over those of unprivileged and unendowed religious persuasions; but they lie under a correlative responsibility to the State, and to every member of the body politic. I am not aware that any sacredness attaches to sermons. If preachers stray beyond the doctrinal limits set by lay lawyers, the Privy Council will see to it ; and, if they think fit to use their pulpits for the promulgation of literary, or historical, or scientific errors, it is not only the right, but the duty, of the humblest layman, who may happen to be better informed, to correct the evil effects of such perversion of the opportunities which the State affords them; and such misuse of the authority which its support lends them. Whatever else it may claim to be, in its relations with the State, the Established Church is a branch of the Civil Service; and, for those who repudiate the ecclesiastical authority of the clergy, they are merely civil servants, as much responsible to the English people for the proper performance of their duties as any others.

The Duke of Argyll tells us that the "work and calling" of the clergy prevent them from "pursuing disputation as others can." I wonder if his Grace ever reads the so-called "religious" newspapers. It is not an occupation which I should commend to any one who wishes to employ his time profitably; but a very short devotion to this exercise will suffice to convince him that the "pursuit of disputation," carried to a degree of acrimony and vehemence unsurpassed in lay controversies, seems to be found quite compatible with the "work and calling" of a remarkably large number of the clergy.

Finally, it appears to me that nothing can be in worse taste than the assumption that a body of English gentlemen can, by any possibility, desire that immunity from criticism which the Duke of

Argyll claims for them. Nothing would be more personally offensive to me than the supposition that I shirked criticism, just or unjust, of any lecture I ever gave. I should be utterly ashamed of myself if, when I stood up as an instructor of others, I had not taken every pains to assure myself of the truth of that which I was about to say; and I should feel myself bound to be even more careful with a popular assembly, who would take me more or less on trust, than with an audience of competent and critical experts.

I decline to assume that the standard of morality, in these matters, is lower among the clergy than it is among scientific men. I refuse to think that the priest who stands up before a congregation, as the minister and interpreter of the Divinity, is less careful in his utterances, less ready to meet adverse comment, than the layman who comes before his audience, as the minister and interpreter of nature. Yet what should we think of the man of science who, when his ignorance or his carelessness was exposed, whined about the want of delicacy of his critics, or pleaded his "work and calling" as a reason for being let alone?

No man, nor any body of men, is good enough, or wise enough, to dispense with the tonic of criticism. Nothing has done more harm to the clergy than the practice, too common among laymen, of regarding them, when in the pulpit, as

a sort of chartered libertines, whose divagations
are not to be taken seriously. And I am well
assured that the distinguished divine, to whom the
sermon is attributed, is the last person who would
desire to avail himself of the dishonouring protection which has been superfluously thrown over
him.

So much for the lecture on propriety. But the
Duke of Argyll, to whom the hortatory style
seems to come naturally, does me the honour to
make my sayings the subjects of a series of other
admonitions, some on philosophical, some on
geological, some on biological topics. I can but
rejoice that the Duke's authority in these matters
is not always employed to show that I am ignorant
of them; on the contrary, I meet with an amount
of agreement, even of approbation, for which I
proffer such gratitude as may be due, even if
that gratitude is sometimes almost overshadowed
by surprise.

I am unfeignedly astonished to find that the
Duke of Argyll, who professes to intervene on
behalf of the preacher, does really, like another
Balaam, bless me altogether in respect of the
main issue.

I denied the justice of the preacher's ascription
to men of science of the doctrine that miracles
are incredible, because they are violations of
natural law; and the Duke of Argyll says that he
believes my "denial to be well-founded. The

preacher was answering an objection which has now been generally abandoned." Either the preacher knew this or he did not know it. It seems to me, as a mere lay teacher, to be a pity that the "great dome of St. Paul's" should have been made to "echo" (if so be that such stentorian effects were really produced) a statement which, admitting the first alternative, was unfair, and, admitting the second, was ignorant.[1]

Having thus sacrificed one half of the preacher's arguments, the Duke of Argyll proceeds to make equally short work with the other half. It appears that he fully accepts my position that the occurrence of those events, which the preacher speaks of as catastrophes, is no evidence of disorder, inasmuch as such catastrophes may be necessary occasional consequences of uniform changes. Whence I conclude, his Grace agrees with me, that the talk about royal laws "wrecking"

[1] The Duke of Argyll speaks of the recent date of the demonstration of the fallacy of the doctrine in question. "Recent" is a relative term, but I may mention that the question is fully discussed in my book on *Hume;* which, if I may believe my publishers, has been read by a good many people since it appeared in 1879. Moreover, I observe, from a note at page 89 of *The Reign of Law,* a work to which I shall have occasion to advert by and by, that the Duke of Argyll draws attention to the circumstance that, so long ago as 1866, the views which I hold on this subject were well known. The Duke, in fact, writing about this time, says, after quoting a phrase of mine : "The question of miracles seems now to be admitted on all hands to be simply a question of evidence." In science, we think that a teacher who ignores views which have been discussed *coram populo* for twenty years, is hardly up to the mark.

ordinary laws may be eloquent metaphor, but is also nonsense.

And now comes a further surprise. After having given these superfluous stabs to the slain body of the preacher's argument, my good ally remarks, with magnificent calmness : " So far, then, the preacher and the professor are at one." " Let them smoke the calumet." By all means : smoke would be the most appropriate symbol of this wonderful attempt to cover a retreat. After all, the Duke has come to bury the preacher, not to praise him ; only he makes the funeral obsequies look as much like a triumphal procession as possible.

So far as the questions between the preacher and myself are concerned, then, I may feel happy. The authority of the Duke of Argyll is ranged on my side. But the Duke has raised a number of other questions, with respect to which I fear I shall have to dispense with his support—nay, even be compelled to differ from him as much, or more, than I have done about his Grace's new rendering of the " benefit of clergy."

In discussing catastrophes, the Duke indulges in statements, partly scientific, partly anecdotic, which appear to me to be somewhat misleading. We are told, to begin with, that Sir Charles Lyell's doctrine respecting the proper mode of interpreting the facts of geology (which is commonly called uniformitarianism) " does not hold

its head quite so high as it once did." That is
great news indeed. But is it true? All I can
say is that I am aware of nothing that has
happened of late that can in any way justify
it; and my opinion is, that the body of Lyell's
doctrine, as laid down in that great work, "The
Principles of Geology," whatever may have happened to its head, is a chief and permanent constituent of the foundations of geological science.

But this question cannot be advantageously discussed, unless we take some pains to discriminate
between the essential part of the uniformitarian
doctrine and its accessories; and it does not
appear that the Duke of Argyll has carried his
studies of geological philosophy so far as this
point. For he defines uniformitarianism to be
the assumption of the "extreme slowness and
perfect continuity of all geological changes."

What "perfect continuity" may mean in this
definition, I am by no means sure; but I can only
imagine that it signifies the absence of any break
in the course of natural order during the millions
of years, the lapse of which is recorded by
geological phenomena.

Is the Duke of Argyll prepared to say that any
geologist of authority, at the present day, believes
that there is the slightest evidence of the occurrence of supernatural intervention, during the
long ages of which the monuments are preserved
to us in the crust of the earth? And if he is not,

in what sense has this part of the uniformitarian doctrine, as he defines it, lowered its pretensions to represent scientific truth ?

As to the "extreme slowness of all geological changes," it is simply a popular error to regard that as, in any wise, a fundamental and necessary dogma of uniformitarianism. It is extremely astonishing to me that any one who has carefully studied Lyell's great work can have so completely failed to appreciate its purport, which yet is "writ large" on the very title-page : "The Principles of Geology, being an attempt to explain the former changes of the earth's surface by reference to causes now in operation." The essence of Lyell's doctrine is here written so that those who run may read ; and it has nothing to do with the quickness or slowness of the past changes of the earth's surface ; except in so far as existing analogous changes may go on slowly, and therefore create a presumption in favour of the slowness of past changes.

With that epigrammatic force which characterises his style, Buffon wrote, nearly a hundred and fifty years ago, in his famous "Théorie de la Terre" : "Pour juger de ce qui est arrivé, et même de ce qui arrivera, nous n'avons qu'à examiner ce qui arrive." The key of the past, as of the future, is to be sought in the present; and, only when known causes of change have been shown to be insufficient, have we any right to have recourse to unknown causes. Geology is as much a historical science as archæology; and I apprehend that all sound historical investigation rests upon this axiom. It underlay all Hutton's work and animated Lyell and Scope in their successful efforts to revolutionise the geology of half a century ago.

There is no antagonism whatever, and there never was, between the belief in the views which had their chief and unwearied advocate in Lyell and the belief in the occurrence of catastrophes. The first edition of Lyell's "Principles," published in 1830, lies before me; and a large part of the first volume is occupied by an account of volcanic, seismic, and diluvial catastrophes which have occurred within the historical period. Moreover, the author, over and over again, expressly draws the attention of his readers to the consistency of catastrophes with his doctrine.

Notwithstanding, therefore, that we have not witnessed within the last three thousand years the devastation by deluge of a large continent, yet, as we may predict the future occurrence of such catastrophes, we are authorised to regard them as part of the present order of nature, and they may be introduced into geological speculations respecting the past, provided that we do not imagine them to have been more frequent or general than we expect them to be in time to come (vol. i. p. 89).

Again :—

If we regard each of the causes separately, which we know to be at present the most instrumental in remodelling the state of the surface, we shall find that we must expect each to be in action for thousands of years, without producing any extensive alterations in the habitable surface, and then to give rise, during a very brief period, to important revolutions (vol. ii. p. 161).[1]

Lyell quarrelled with the catastrophists then, by no means because they assumed that catastrophes occur and have occurred, but because they had got into the habit of calling on their god Catastrophe to help them, when they ought to have been putting their shoulders to the wheel of observation of the present course of nature, in order to help themselves out of their difficulties. And geological science has become what it is, chiefly because geologists have gradually accepted Lyell's doctrine and followed his precepts.

So far as I know anything about the matter, there is nothing that can be called proof, that the causes of geological phenomena operated more intensely or more rapidly, at any time between the older tertiary and the oldest palæozoic epochs than they have done between the older tertiary epoch and the present day. And if that is so, uniformitarianism, even as limited by Lyell,[2] has no

[1] See also vol. i. p. 460. In the ninth edition (1853), published twenty-three years after the first, Lyell deprives even the most careless reader of any excuse for misunderstanding him: "So in regard to subterranean movements, the theory of the perpetual uniformity of the force which they exert on the earth-crust is quite consistent with the admission of their alternate development and suspension for indefinite periods within limited geographical areas " (p. 187).

[2] A great many years ago (Presidential Address to the Geological Society, 1869) I ventured to indicate that which seemed to me to be the weak point, not in the fundamental principles of uniformitarianism, but in uniformitarianism as taught by Lyell. It lay, to my mind, in the refusal by Hutton, and in a

call to lower its crest. But if the facts were otherwise, the position Lyell took up remains impregnable.
He did not say that the geological operations of
nature were never more rapid, or more vast, than
they are now ; what he did maintain is the very
different proposition that there is no good evidence
of anything of the kind. And that proposition
has not yet been shown to be incorrect.

I owe more than I can tell to the careful study
of the "Principles of Geology" in my young
days; and, long before the year 1856, my mind
was familiar with the truth that "the doctrine of
uniformity is not incompatible with great and
sudden changes," which, as I have shown, is
taught *totidem verbis* in that work. Even had it
been possible for me to shut my eyes to the sense
of what I had read in the "Principles," Whewell's
"Philosophy of the Inductive Sciences," published
in 1840, a work with which I was also tolerably
familiar, must have opened them. For the
always acute, if not always profound, author, in
arguing against Lyell's uniformitarianism, ex-

---

less degree by Lyell, to look beyond the limits of the time
recorded by the stratified rocks. I said : "This attempt to
limit, at a particular point, the progress of inductive and de-
ductive reasoning from the things which are to the things which
were—this faithlessness to its own logic, seems to me to have
cost uniformitarianism the place as the permanent form of geological speculation which it might otherwise have held" (*Lay
Sermons*, p. 260). The context shows that "uniformitarianism"
here means that doctrine, as limited in application by Hutton
and Lyell, and that what I mean by "evolutionism" is consistent and thoroughgoing uniformitarianism.

pressly points out that it does not in any way contravene the occurrence of catastrophes.

With regard to such occurrences [earthquakes, deluges, etc.], terrible as they appear at the time, they may not much affect the average rate of change : there may be a *cycle*, though an irregular one, of rapid and slow change : and if such cycles go on succeeding each other, we may still call the order of nature uniform, notwithstanding the periods of violence which it involves.[1]

The reader who has followed me through this brief chapter of the history of geological philosophy will probably find the following passage in the paper of the Duke of Argyll to be not a little remarkable :—

Many years ago, when I had the honour of being President of the British Association,[2] I ventured to point out, in the presence and in the hearing of that most distinguished man [Sir C. Lyell] that the doctrine of uniformity was not incompatible with great and sudden changes, since cycles of these and other cycles of comparative rest might well be constituent parts of that uniformity which he asserted. Lyell did not object to this extended interpretation of his own doctrine, and indeed expressed to me his entire concurrence.

I should think he did; for, as I have shown, there was nothing in it that Lyell himself had not said, six-and-twenty years before, and enforced, three years before; and it is almost verbally identical with the view of uniformitarianism taken by Whewell, sixteen years before, in a work with which, one would think, that any one who

[1] *Philosophy of the Inductive Sciences*, vol. i. p. 670. New edition, 1847.   [2] At Glasgow in 1856.

undertakes to discuss the philosophy of science should be familiar.

Thirty years have elapsed since the beginner of 1856 persuaded himself that he enlightened the foremost geologist of his time, and one of the most acute and far-seeing men of science of any time, as to the scope of the doctrines which the veteran philosopher had grown gray in promulgating; and the Duke of Argyll's acquaintance with the literature of geology has not, even now, become sufficiently profound to dissipate that pleasant delusion.

If the Duke of Argyll's guidance in that branch of physical science, with which alone he has given evidence of any practical acquaintance, is thus unsafe, I may breathe more freely in setting my opinion against the authoritative deliverances of his Grace about matters which lie outside the province of geology.

And here the Duke's paper offers me such a wealth of opportunities that choice becomes embarrassing. I must bear in mind the good old adage, " Non multa sed multum." Tempting as it would be to follow the Duke through his labyrinthine misunderstandings of the ordinary terminology of philosophy, and to comment on the curious unintelligibility which hangs about his frequent outpourings of fervid language, limits of space oblige me to restrict myself to those points, the discussion of which may help to enlighten the public in respect of matters of more importance than the competence of my Mentor for the task which he has undertaken.

I am not sure when the employment of the word Law, in the sense in which we speak of laws of nature, commenced, but examples of it may be found. in the works of Bacon, Descartes, and Spinoza. Bacon employs "Law" as the equivalent of "Form," and I am inclined to think that he may be responsible for a good deal of the confusion that has subsequently arisen; but I am not aware that the term is used by other authorities, in the seventeenth and eighteenth centuries, in any other sense than that of "rule" or "definite order" of the coexistence of things or succession of events in nature. Descartes speaks of "règles, que je nomme les lois de la nature." Leibnitz says "loi ou règle générale," as if he considered the terms interchangeable.

The Duke of Argyll, however, affirms that the "law of gravitation" as put forth by Newton was something more than the statement of an observed order. He admits that Kepler's three laws "were an observed order of facts and nothing more." As to the law of gravitation, "it contains an element which Kepler's laws did not contain, even an element of causation, the recognition of which belongs to a higher category of intellectual conceptions than that which is concerned in the mere observation and record of separate and apparently

assertion that the notion of a body acting where
it is not, is one that no competent thinker could
entertain, is antagonistic to the whole current
conception of attractive and repulsive forces, and
therefore of "the attractive force of gravitation."
What, then, was that labour of unsurpassed magnitude and excellence and of immortal influence
which Newton did perform ?   In the first place,
Newton defined the laws, rules, or observed order
of the phenomena of motion, which come under
our daily observation, with greater precision than
had been before attained ; and, by following out,
with marvellous power and subtlety, the mathematical consequences of these rules, he almost
created the modern science of pure mechanics.
In the second place, applying exactly the same
method to the explication of the facts of astronomy as that which was applied a century and a
half later to the facts of geology by Lyell, he set
himself to solve the following problem. Assuming
that all bodies, free to move, tend to approach
one another as the earth and the bodies on it do ;
assuming that the strength of that tendency is
directly as the mass and inversely as the squares
of the distances ; assuming that the laws of
motion, determined for terrestrial bodies, hold
good throughout the universe ; assuming that
the planets and their satellites were created and
placed at their observed mean distances, and that
each received a certain impulse from the Creator;

will the form of the orbits, the varying rates of
motion of the planets, and the ratio between
those rates and their distances from the sun,
which must follow by mathematical reasoning
from these premisses, agree with the order of
facts determined by Kepler and others, or not ?

Newton, employing mathematical methods
which are the admiration of adepts, but which
no one but himself appears to have been able
to use with ease, not only answered this question
in the affirmative, but stayed not his constructive
genius before it had founded modern physical
astronomy.

The historians of mechanical and of astronomical science appear to be agreed that he was the
first person who clearly and distinctly put forth
the hypothesis that the phenomena comprehended
under the general name of "gravity" follow the
same order throughout the universe, and that all
material bodies exhibit these phenomena; so that,
in this sense, the idea of universal gravitation
may, doubtless, be properly ascribed to him.

Newton proved that the laws of Kepler were
particular consequences of the laws of motion
and the law of gravitation—in other words, the
reason of the first lay in the two latter. But to
talk of the law of gravitation alone as the reason
of Kepler's laws, and still more as standing in
any causal relation to Kepler's laws, is simply a
misuse of language. It would really be interest-

acting as a cause in a way quite in accordance with the Duke of Argyll's conception of it. In fact, in the mind of the author of the "Vestiges," "laws" are existences intermediate between the Creator and His works, like the "ideas" of the Platonisers or the Logos of the Alexandrians.[1] I may cite a passage which is quite in the vein of Philo :—

We have seen powerful evidences that the construction of this globe and its associates ; and, inferentially, that of all the other globes in space, was the result, not of any immediate or personal exertion on the part of the Deity, but of natural laws which are the expression of His will. What is to hinder our supposing that the organic creation is also a result of natural laws which are in like manner an expression of His will ? (p. 154, 1st edition).

And creation "operating by law" is constantly cited as relieving the Creator from trouble about insignificant details.

I am perplexed to picture to myself the state of mind which accepts these verbal juggleries. It is intelligible that the Creator should operate according to such rules as he might think fit to lay down for himself (and therefore according to law) ; but that would leave the operation of his will just as much a direct personal act as it would be under any other circumstances. I can also understand that (as in Leibnitz's caricature of Newton's views) the Creator might have made

---

[1] The author recognises this in his *Explanations*.

the cosmical machine, and, after setting it going, have left it to itself till it needed repair. But then, by the supposition, his personal responsibility would have been involved in all that it did; just as much as a dynamiter is responsible for what happens, when he has set his machine going and left it to explode.

The only hypothesis which gives a sort of mad consistency to the Vestigiarian's views is the supposition that laws are a kind of angels or demiurgoi, who, being supplied with the Great Architect's plan, were permitted to settle the details among themselves. Accepting this doctrine, the conception of royal laws and plebeian laws, and of those more than Homeric contests in which the big laws "wreck" the little ones, becomes quite intelligible. And, in fact, the honour of the paternity of those remarkable ideas which come into full flower in the preacher's discourse, must, so far as my imperfect knowledge goes, be attributed to the author of the Vestiges."

But the author of the "Vestiges" is not the only writer who is responsible for the current pseudo-scientific mystifications which hang about the term "law." When I wrote my paper about "Scientific and Pseudo-Scientific Realism," I had not read a work by the Duke of Argyll, "The Reign of Law," which, I believe, has enjoyed, possibly still enjoys, a widespread popularity. But the vivacity of the Duke's attack led me to

think it possible that criticisms directed elsewhere might have come home to him. And, in
fact, I find that the second chapter of the work in
question, which is entitled " Law; its definitions,"
is, from my point of view, a sort of "summa" of
pseudo-scientific philosophy. It will be worth
while to examine it in some detail.

In the first place, it is to be noted that the
author of the " Reign of Law " admits that " law,"
in many cases, means nothing more than the
statement of the order in which facts occur, or, as
he says, "an observed order of facts " (p. 66).
But his appreciation of the value of accuracy of
expression does not hinder him from adding,
almost in the same breath, "In this sense the
laws of nature are simply those facts of nature
which recur according to rule " (p. 66). Thus
" laws," which were rightly said to be the statement of an order of facts in one paragraph, are
declared to be the facts themselves in the next.

We are next told that, though it may be
customary and permissible to use "law " in the
sense of a statement of the order of facts, this is
a low use of the word ; and, indeed, two pages
farther on, the writer, flatly contradicting himself,
altogether denies its admissibility.

An observed order of facts, to be entitled to the rank of a law,
must be an order so constant and uniform as to indicate necessity,
and necessity can only arise out of the action of some compelling
force (p. 68).

This is undoubtedly one of the most singular propositions that I have ever met with in a professedly scientific work, and its rarity is embellished by another direct self-contradiction which it implies. For on the preceding page (67), when the Duke of Argyll is speaking of the laws of Kepler, which he admits to be laws, and which are types of that which men of science understand by "laws," he says that they are "simply and purely an order of facts." Moreover, he adds: "A very large proportion of the laws of every science are laws of this kind and in this sense."

If, according to the Duke of Argyll's admission, law is understood, in this sense, thus widely and constantly by scientific authorities, where is the justification for his unqualified assertion that such statements of the observed order of facts are not "entitled to the rank" of laws?

But let us examine the consequences of the really interesting proposition I have just quoted. I presume that it is a law of nature that "a straight line is the shortest distance between two points." This law affirms the constant association of a certain fact of form with a certain fact of dimension. Whether the notion of necessity which attaches to it has an *a priori* or an *a posteriori* origin is a question not relevant to the present discussion. But I would beg to be informed, if it is necessary, where is the "compelling force" out of which the necessity arises ; and further, if it is not necessary, whether it loses the character of a law of nature ?

I take it to be a law of nature, based on unexceptionable evidence, that the mass of matter remains unchanged, whatever chemical or other modifications it may undergo. This law is one of the foundations of chemistry. But it is by no means necessary. It is quite possible to imagine that the mass of matter should vary according to circumstances, as we know its weight does. Moreover, the determination of the " force " which makes mass constant (if there is any intelligibility in that form of words) would not, so far as I can see, confer any more validity on the law than it has now.

There is a law of nature, so well vouched by experience, that all mankind, from pure logicians in search of examples to parish sextons in search of fees, confide in it. This is the law that " all men are mortal." It is simply a statement of the observed order of facts that all men sooner or later die. I am not acquainted with any law of nature which is more " constant and uniform " than this. But will any one tell me that death is " necessary "? Certainly there is no *à priori* necessity in the case, for various men have been imagined to be immortal. And I should be glad to be informed of any " necessity " that can be deduced from biological considerations. It is quite conceivable, as has recently been pointed out, that some of the lowest forms of life may be immortal, after a fashion. However this may be, I would further ask, supposing "all men are mortal" to be a real law of nature, where and what is that to which, with any propriety, the title of "compelling force" of the law can be given ?

On page 69, the Duke of Argyll asserts that the law of gravitation "is a law in the sense, not merely of a rule, but of a cause." But this revival of the teaching of the "Vestiges" has already been examined and disposed of; and when the Duke of Argyll states that the "observed order" which Kepler had discovered was simply a necessary consequence of the force of "gravitation," I need not recapitulate the evidence which proves such a statement to be wholly fallacious. But it may be useful to say, once more, that, at this present moment, nobody knows anything about the existence of a "force" of gravitation apart from the fact; that Newton declared the ordinary notion of such force to be inconceivable; that various attempts have been made to account for the order of facts we call gravitation, without recourse to the notion of attractive force ; that, if such a force exists, it is utterly incompetent to account for Kepler's laws, without taking into the reckoning a great number of other considerations ; and, finally, that all we know about the "force"

of gravitation, or any other so-called "force," is
that it is a name for the hypothetical cause of an
observed order of facts.

Thus, when the Duke of Argyll says : "Force, ascertained according to some measure of its operation
—this is indeed one of the definitions, but only
one, of a scientific law " (p. 71) I reply that it is a
definition which must be repudiated by every one
who possesses an adequate acquaintance with
either the facts, or the philosophy, of science, and be
relegated to the limbo of pseudo-scientific fallacies.
If the human mind had never entertained this
notion of "force," nay, if it substituted bare in-
variable succession for the ordinary notion of
causation, the idea of law, as the expression of a
constantly-observed order, which generates a corresponding intensity of expectation in our minds,
would have exactly the same value, and play its
part in real science, exactly as it does now.

It is needless to extend further the present
excursus on the origin and history of modern
pseudo-science. Under such high patronage as
it has enjoyed, it has grown and flourished until,
nowadays, it is becoming somewhat rampant.
It has its weekly "Ephemerides," in which every
new pseudo-scientific mare's-nest is hailed and
belauded with the unconscious unfairness of
ignorance ; and an army of "reconcilers," enlisted
in its service, whose business seems to be to mix
the black of dogma and the white of science into

I 2

the neutral tint of what they call liberal
theology.

I remember that, not long after the publication of
the "Vestiges," a shrewd and sarcastic countryman
of the author defined it as "cauld kail made het
again." A cynic might find amusement in the
reflection that, at the present time, the principles
and the methods of the much-vilified Vestigiarian
are being "made het again"; and are not only
"echoed by the dome of St. Paul's," but thundered
from the castle of Inverary. But my turn of
mind is not cynical, and I can but regret the
waste of time and energy bestowed on the endeavour to deal with the most difficult problems
of science, by those who have neither undergone
the discipline, nor possess the information, which
are indispensable to the successful issue of such
an enterprise.

I have already had occasion to remark that the
Duke of Argyll's views of the conduct of controversy are different from mine; and this much-
to-be lamented discrepancy becomes yet more
accentuated when the Duke reaches biological
topics. Anything that was good enough for Sir
Charles Lyell, in his department of study, is certainly good enough for me in mine; and I by no
means demur to being pedagogically instructed
about a variety of matters with which it has been
the business of my life to try to acquaint myself.
But the Duke of Argyll is not content with

favouring me with his opinions about my own business; he also answers for mine; and, at that point, really the worm must turn. I am told that "no one knows better than Professor Huxley" a variety of things which I really do not know; and I am said to be a disciple of that "Positive Philosophy" which I have, over and over again, publicly repudiated in language which is certainly not lacking in intelligibility, whatever may be its other defects.

I am told that I have been amusing myself with a "metaphysical exercitation or logomachy" (may I remark incidentally that these are not quite convertible terms?), when, to the best of my belief, I have been trying to expose a process of mystification, based upon the use of scientific language by writers who exhibit no sign of scientific training, of accurate scientific knowledge, or of clear ideas respecting the philosophy of science, which is doing very serious harm to the public. Naturally enough, they take the lion's skin of scientific phraseology for evidence that the voice which issues from beneath it is the voice of science, and I desire to relieve them from the consequences of their error.

The Duke of Argyll asks, apparently with sorrow that it should be his duty to subject me to reproof—

What shall we say of a philosophy which confounds the organic with the inorganic, and, refusing to take note of a difference so profound, assumes to explain under one common abstraction, the movements due to gravitation and the movements due to the mind of man ?

To which I may fitly reply by another question : What shall we say to a controversialist who attributes to the subject of his attack opinions which are notoriously not his; and expresses himself in such a manner that it is obvious he is unacquainted with even the rudiments of that knowledge which is necessary to the discussion into which he has rushed ?

What line of my writing can the Duke of Argyll produce which confounds the organic with the inorganic ?

As to the latter half of the paragraph, I have to confess a doubt whether it has any definite meaning.   But I imagine that the Duke is alluding to my assertion that the law of gravitation is nowise " suspended " or " defied " when a man lifts his arm; but that, under such circumstances, part of the store of energy in the universe operates on the arm at a mechanical advantage as against the operation of another part.   I was simple enough to think that no one who had as much knowledge of physiology as is to be found in an elementary primer, or who had ever heard of the greatest physical generalisation of modern times—the doctrine of the conservation of energy—would dream of doubting my statement; and I was further simple enough to think that no one who lacked these qualifications would feel tempted to charge me with error. It appears that my simplicity is greater than my powers of imagination.

The Duke of Argyll may not be aware of the fact, but it is nevertheless true, that when a man's arm is raised, in sequence to that state of consciousness we call a volition, the volition is not the immediate cause of the elevation of the arm. On the contrary, that operation is effected by a certain change of form, technically known as " contraction " in sundry masses of flesh, technically known as muscles, which are fixed to the bones of the shoulder in such a manner that, if these muscles contract, they must raise the arm. Now each of these muscles is a machine comparable, in a certain sense, to one of the donkey-engines of a steamship, but more complete, inasmuch as the source of its ability to change its form, or contract, lies within itself. Every time that, by contracting, the muscle does work, such as that involved in raising the arm, more or less of the material which it contains is used up, just as more or less of the fuel of a steam-engine is used up, when it does work. And I do not think there is a doubt in the mind of any competent physicist, or physiologist, that the work done in lifting the weight of the arm is the mechanical equivalent of a certain proportion of the energy set free by the molecular changes which take place in the muscle. It is further a tolerably well-based belief that this, and all other

of directly affecting the motion of even the smallest conceivable molecule of matter ? Is such a
thing even conceivable ? If we answer these
questions in the negative, it follows that volition
may be a sign, but cannot be a cause, of bodily
motion. If we answer them in the affirmative, then
states of consciousness become undistinguishable
from material things ; for it is the essential nature
of matter to be the vehicle or substratum of
mechanical energy.

There is nothing new in all this. I have
merely put into modern language the issue
raised by Descartes more than two centuries ago.
The philosophies of the Occasionalists, of Spinoza,
of Malebranche, of modern idealism and modern
materialism, have all grown out of the controversies which Cartesianism evoked. Of all this
the pseudo-science of the present time appears to
be unconscious; otherwise it would hardly content
itself with "making het again" the pseudo-science
of the past.

In the course of these observations I have
already had occasion to express my appreciation
of the copious and perfervid eloquence which
enriches the Duke of Argyll's pages. I am
almost ashamed that a constitutional insensibility
to the Sirenian charms of rhetoric has permitted
me, in wandering through these flowery meads, to
be attracted, almost exclusively, to the bare
places of fallacy and the stony grounds of deficient

information, which are disguised, though not concealed, by these floral decorations. But, in his concluding sentences, the Duke soars into a Tyrtæan strain which roused even my dull soul.

It was high time, indeed, that some revolt should be raised against that Reign of Terror which had come to be established in the scientific world under the abuse of a great name. Professor Huxley has not joined this revolt openly, for as yet, indeed, it is only beginning to raise its head. But more than once—and very lately—he has uttered a warning voice against the shallow dogmatism that has provoked it. The time is coming when that revolt will be carried further. Higher interpretations will be established. Unless I am much mistaken, they are already coming in sight (p. 339).

I have been living very much out of the world for the last two or three years, and when I read this denunciatory outburst, as of one filled with the spirit of prophecy, I said to myself, "Mercy upon us, what has happened? Can it be that X. and Y. (it would be wrong to mention the names of the vigorous young friends which occurred to me) are playing Danton and Robespierre; and that a guillotine is erected in the courtyard of Burlington House for the benefit of all anti-Darwinian Fellows of the Royal Society? Where are the secret conspirators against this tyranny, whom I am supposed to favour, and yet not have the courage to join openly? And to think of my poor oppressed friend, Mr. Herbert Spencer, ' compelled to speak with bated breath' (p. 338) certainly for the first time in my thirty-odd years' acquaintance with him!" My alarm and horror at the supposition that, while I had been fiddling (or at any rate physicking), my beloved Rome had been burning, in this fashion, may be imagined.

I am sure the Duke of Argyll will be glad to hear that the anxiety he created was of extremely short duration. It is my privilege to have access to the best sources of information, and nobody in the scientific world can tell me anything about either the "Reign of Terror" or "the Revolt." In fact, the scientific world laughs most indecorously at the notion of the existence of either; and some are so lost to the sense of the scientific dignity, that they descend to the use of transatlantic slang, and call it a "bogus scare." As to my friend Mr. Herbert Spencer, I have every reason to know that, in the "Factors of Organic Evolution," he has said exactly what was in his mind, without any particular deference to the opinions of the person whom he is pleased to regard as his most dangerous critic and Devil's Advocate-General, and still less of any one else.

I do not know whether the Duke of Argyll pictures himself as the Tallien of this imaginary revolt against a no less imaginary Reign of Terror. But if so, I most respectfully but firmly decline to join his forces. It is only a few weeks since I happened to read over again the first article which I ever wrote (now twenty-seven years ago)

on the " Origin of Species," and I found nothing
that I wished to modify in the opinions that are
there expressed, though the subsequent vast
accumulation of evidence in favour of Mr. Darwin's views would give me much to add. As is
the case with all new doctrines, so with that of
Evolution, the enthusiasm of advocates has sometimes tended to degenerate into fanaticism; and
mere speculation has, at times, threatened to
shoot beyond its legitimate bounds. I have
occasionally thought it wise to warn the more
adventurous spirits among us against these
dangers; in sufficiently plain language ; and I
have sometimes jestingly said that I expected,
if I lived long enough, to be looked on as a
reactionary by some of my more ardent friends.
But nothing short of midsummer madness can
account for the fiction that I am waiting till it is
safe to join openly a revolt, hatched by some
person or persons unknown, against an intellectual
movement with which I am in the most entire
and hearty sympathy. It is a great many years
since, at the outset of my career, I had to think
seriously what life had to offer that was worth
having. I came to the conclusion that the chief
good, for me, was freedom to learn, think, and say
what I pleased, when I pleased. I have acted on
that conviction, and have availed myself of the
"rara temporum felicitas ubi sentire quæ velis, et
quæ sentias dicere licet," which is now enjoyable,

to the best of my ability; and though strongly, and perhaps wisely, warned that I should probably come to grief, I am entirely satisfied with the results of the line of action I have adopted.

My career is at an end. I have

Warmed both hands before the fire of life;

and nothing is left me, before I depart, but to help, or at any rate to abstain from hindering, the younger generation of men of science in doing better service to the cause we have at heart than I have been able to render.

And yet, forsooth, I am supposed to be waiting for the signal of " revolt," which some fiery spirits among these young men are to raise before I dare express my real opinions concerning questions about which we older men had to fight, in the teeth of fierce public opposition and obloquy—of something which might almost justify even the grandiloquent epithet of a Reign of Terror—before our excellent successors had left school.

It would appear that the spirit of pseudo-science has impregnated even the imagination of the Duke of Argyll. The scientific imagination always restrains itself within the limits of probability.

IV

AN EPISCOPAL TRILOGY

[1887]

IF there is any truth in the old adage that a
burnt child dreads the fire, I ought to be very
loath to touch a sermon, while the memory of what
befell me on a recent occasion, possibly not yet
forgotten by the readers of the *Nineteenth Century*,
is uneffaced. But I suppose that even the distinguished censor of that unheard-of audacity to
which not even the newspaper report of a sermon
is sacred, can hardly regard a man of science as
either indelicate or presumptuous, if he ventures
to offer some comments upon three discourses,
specially addressed to the great assemblage of
men of science which recently gathered at
Manchester, by three bishops of the State Church.
On my return to England not long ago, I found a
pamphlet [1] containing a version, which I presume

<hr />

[1] *The Advance of Science.* Three sermons preached in Manchester Cathedral on Sunday, September 4, 1887, during the

to be authorised, of these sermons, among the huge mass of letters and papers which had accumulated during two months' absence; and I have read them not only with attentive interest, but with a feeling of satisfaction which is quite new to me as a result of hearing, or reading, sermons. These excellent discourses, in fact, appear to me to signalise a new departure in the course adopted by theology towards science, and to indicate the possibility of bringing about an honourable *modus vivendi* between the two. How far the three bishops speak as accredited representatives of the Church is a question to be considered by and by. Most assuredly, I am not authorised to represent any one but myself. But I suppose that there must be a good many people in the Church of the bishops' way of thinking; and I have reason to believe that, in the ranks of science, there are a good many persons who, more or less, share my views. And it is to these sensible people on both sides, as the bishops and I must needs think those who agree with us, that my present observations are addressed. They will probably be astonished to learn how insignificant, in principle, their differences are.

It is impossible to read the discourses of the three prelates without being impressed by the

meeting of the British Association for the Advancement of Science, by the Bishop of Carlisle, the Bishop of Bedford, and the Bishop of Manchester.

knowledge which they display, and by the spirit
of equity, I might say of generosity, towards
science which pervades them. There is no trace
of that tacit or open assumption that the rejection
of theological dogmas, on scientific grounds, is due
to moral perversity, which is the ordinary note of
ecclesiastical homilies on this subject, and which
makes them look so supremely silly to men whose
lives have been spent in wrestling with these
questions. There is no attempt to hide away real
stumbling-blocks under rhetorical stucco; no resort
to the *tu quoque* device of setting scientific blunders against theological errors; no suggestion that
an honest man may keep contradictory beliefs in
separate pockets of his brain; no question that the
method of scientific investigation is valid, whatever the results to which it may lead; and that the
search after truth, and truth only, ennobles the
searcher and leaves no doubt that his life, at any
rate, is worth living. The Bishop of Carlisle
declares himself pledged to the belief that "the
advancement of science, the progress of human
knowledge, is in itself a worthy aim of the greatest
effort of the greatest minds."

How often was it my fate, a quarter of a century
ago, to see the whole artillery of the pulpit brought
to bear upon the doctrine of evolution and its sup-
porters! Any one unaccustomed to the amenities
of ecclesiastical controversy would have thought
we were too wicked to be permitted to live. But

let us hear the Bishop of Bedford. After a perfectly frank statement of the doctrine of evolution and some of its obvious consequences, that learned prelate pleads, with all earnestness, against

a hasty denunciation of what *may* be proved to have at least some elements of truth in it, a contemptuous rejection of theories which we *may* some day learn to accept as freely and with as little sense of inconsistency with God's word as we now accept the theory of the earth's motion round the sun, or the long duration of the geological epochs (p. 28).

I do not see that the most convinced evolutionist could ask any one, whether cleric or layman, to say more than this; in fact, I do not think that any one has a right to say more, with respect to any question about which two opinions can be held, than that his mind is perfectly open to the force of evidence.

There is another portion of the Bishop of Bedford's sermon which I think will be warmly appreciated by all honest and clear-headed men. He repudiates the views of those who say that theology and science

occupy wholly different spheres, and need in no way intermeddle with each other. They revolve, as it were, in different planes, and so never meet. Thus we may pursue scientific studies with the utmost freedom and, at the same time, may pay the most reverent regard to theology, having no fears of collision, because allowing no points of contact (p. 29).

Surely every unsophisticated mind will heartily

concur with the Bishop's remark upon this convenient refuge for the descendants of Mr. Facing-
both-ways. " I have never been able to understand this position, though I have often seen it
assumed." Nor can any demurrer be sustained
when the Bishop proceeds to point out that there
are, and must be, various points of contact between
theological and natural science, and therefore that
it is foolish to ignore or deny the existence of as
many dangers of collision.

Finally, the Bishop of Manchester freely admits
the force of the objections which have been raised,
on scientific grounds, to prayer, and attempts to
turn them by arguing that the proper objects of
prayer are not physical but spiritual. He tells us
that natural accidents and moral misfortunes are
not to be taken for moral judgments of God; he
admits the propriety of the application of scientific
methods to the investigation of the origin and
growth of religions; and he is as ready to recognise
the process of evolution there, as in the physical
world. Mark the following striking passage :—

And how utterly all the common objections to Divine revelation vanish away when they are set in the light of this theory of
a spiritual progression. Are we reminded that there prevailed,
in those earlier days, views of the nature of God and man, of
human life and Divine Providence, which we now find to be
untenable ? *That*, we answer, is precisely what the theory of
development presupposes. If early views of religion and morality had not been imperfect, where had been the development ?
If symbolical visions and mythical creations had found no place

in the early Oriental expression of Divine truth, where had been the development ? The sufficient answer to ninety-nine out of a hundred of the ordinary objections to the Bible, as the record of a divine education of our race, is asked in that one word— development. And to what are we indebted for that potent word, which, as with the wand of a magician, has at the same moment so completely transformed our knowledge and dispelled our difficulties ? To modern science, resolutely pursuing its search for truth in spite of popular obloquy and—alas ! that one should have to say it—in spite too often of theological denunciation (p. 53).

Apart from its general importance, I read this remarkable statement with the more pleasure, since, however imperfectly I may have endeavoured to illustrate the evolution of theology in a paper published in the *Nineteenth Century* last year,[1] it seems to me that in principle, at any rate, I may hereafter claim high theological sanction for the views there set forth.

If theologians are henceforward prepared to recognise the authority of secular science in the manner and to the extent indicated in the Manchester trilogy; if the distinguished prelates who offer these terms are really plenipotentiaries, then, so far as I may presume to speak on such a matter, there will be no difficulty about concluding a perpetual treaty of peace, and indeed of alliance, between the high contracting powers, whose history has hitherto been little more than a record of continual warfare. But if the great Chancellor's

---

[1] Reprinted in Vol. IV. of this collection.

maxim, "Do ut des," is to form the basis of negotiation, I am afraid that secular science will be ruined; for it seems to me that theology, under the generous impulse of a sudden conversion, has given all that she hath; and indeed, on one point, has surrendered more than can reasonably be asked.

I suppose I must be prepared to face the reproach which attaches to those who criticise a gift, if I venture to observe that I do not think that the Bishop of Manchester need have been so much alarmed, as he evidently has been, by the objections which have often been raised to prayer, on the ground that a belief in the efficacy of prayer is inconsistent with a belief in the constancy of the order of nature.

The Bishop appears to admit that there is an antagonism between the "regular economy of nature" and the "regular economy of prayer" (p. 39), and that "prayers for the interruption of God's natural order" are of "doubtful validity" (p. 42). It appears to me that the Bishop's difficulty simply adds another example to those which I have several times insisted upon in the pages of this Review and elsewhere, of the mischief which has been done, and is being done, by a mistaken apprehension of the real meaning of "natural order" and "law of nature."

May I, therefore, be permitted to repeat, once more, that the statements denoted by these terms have no greater value or cogency than such as may

which they are strong enough to modify or control;
and who is capable of being moved by appeals
such as men make to one another. This belief
does not even involve theism ; for our earth is an
insignificant particle of the solar system, while the
solar system is hardly worth speaking of in relation
to the All; and, for anything that can be proved
to the contrary, there may be beings endowed
with full powers over our system, yet, practically,
as insignificant as ourselves in relation to the
universe. If any one pleases, therefore, to give
unrestrained liberty to his fancy, he may plead
analogy in favour of the dream that there may be,
somewhere, a finite being, or beings, who can play
with the solar system as a child plays with a toy;
and that such being may be willing to do anything
which he is properly supplicated to do. For we
are not justified in saying that it is impossible for
beings having the nature of men, only vastly more
powerful, to exist ; and if they do exist, they may
act as and when we ask them to do so, just as our
brother men act. As a matter of fact, the great
mass of the human race has believed, and still
believes, in such beings, under the various names
of fairies, gnomes, angels, and demons. Certainly
I do not lack faith in the constancy of natural
order. But I am not less convinced that if I were
to ask the Bishop of Manchester to do me a kind-
ness which lay within his power, he would do it.
And I am unable to see that his action on my
request involves any violation of the order of

nature. On the contrary, as I have not the honour to know the Bishop personally, my action would be based upon my faith in that "law of nature," or generalisation from experience, which tells me that, as a rule, men who occupy the Bishop's position are kindly and courteous. How is the case altered if my request is preferred to some imaginary superior being, or to the Most High being, who, by the supposition, is able to arrest disease, or make the sun stand still in the heavens, just as easily as I can stop my watch, or make it indicate any hour that pleases me?

I repeat that it is not upon any *a priori* considerations that objections, either to the supposed efficacy of prayer in modifying the course of events, or to the supposed occurrence of miracles, can be scientifically based. The real objection, and, to my mind, the fatal objection, to both these suppositions, is the inadequacy of the evidence to prove any given case of such occurrences which has been adduced. It is a canon of common sense, to say nothing of science, that the more improbable a supposed occurrence, the more cogent ought to be the evidence in its favour. I have looked somewhat carefully into the subject, and I am unable to find in the records of any miraculous event evidence which even approximates to the fulfilment of this requirement.

But, in the case of prayer, the Bishop points out a most just and necessary distinction between its effect on the course of nature, outside ourselves, and its effect within the region of the supplicator's mind.

It is a "law of nature," verifiable by everyday experience, that our already formed convictions, our strong desires, our intent occupation with particular ideas, modify our mental operations to a most marvellous extent, and produce enduring changes in the direction and in the intensity of our intellectual and moral activities. Men can intoxicate themselves with ideas as effectually as with alcohol or with bang, and produce, by dint of intense thinking, mental conditions hardly distinguishable from monomania. Demoniac possession is mythical; but the faculty of being possessed, more or less completely, by an idea is probably the fundamental condition of what is called genius, whether it show itself in the saint, the artist, or the man of science. One calls it faith, another calls it inspiration, a third calls it insight; but the "intending of the mind," to borrow Newton's well-known phrase, the concentration of all the rays of intellectual energy on some one point, until it glows and colours the whole cast of thought with its peculiar light, is common to all.

I take it that the Bishop of Manchester has psychological science with him when he insists upon the subjective efficacy of prayer in faith, and on the seemingly miraculous effects which such "intending of the mind" upon religious and moral ideals may have upon character and happiness. Scientific faith, at present, takes it no further than the prayer which Ajax offered; but that petition is continually granted.

Whatever points of detail may yet remain open for discussion, however, I repeat the opinion I have already expressed, that the Manchester sermons concede all that science, has an indisputable right, or any pressing need, to ask, and that not grudgingly but generously; and, if the three bishops of 1887 carry the Church with them, I think they will have as good title to the permanent gratitude of posterity as the famous seven who went to the Tower in defence of the Church two hundred years ago.

Will their brethren follow their just and prudent guidance? I have no such acquaintance with the currents of ecclesiastical opinion as would justify me in even hazarding a guess on such a difficult topic. But some recent omens are hardly favourable. There seems to be an impression abroad—I do not desire to give any countenance to it—that I am fond of reading sermons. From time to time, unknown correspondents—some apparently animated by the charitable desire to promote my conversion, and others unmistakably anxious to spur me to the expression of wrathful antagonism—favour me with reports or copies of such productions.

I found one of the latter category among the accumulated arrears to which I have already referred.

It is a full, and apparently accurate, report of a discourse by a person of no less ecclesiastical rank than the three authors of the sermons I have hitherto been considering; but who he is, and where or when the sermon was preached, are secrets which wild horses shall not tear from me, lest I fall again under high censure for attacking a clergyman. Only if the editor of this Review thinks it his duty to have independent evidence that the sermon has a real existence, will I, in the strictest confidence, communicate it to him.

The preacher, in this case, is of a very different mind from the three bishops—and this mind is different in quality, different in spirit, and different in contents. He discourses on the *a priori* objections to miracles, apparently without being aware, in spite of all the discussions of the last seven or eight years, that he is doing battle with a shadow.

I trust I do not misrepresent the Bishop of Manchester in saying that the essence of his remarkable discourse is the insistence upon the "supreme importance of the purely spiritual in our faith," and of the relative, if not absolute, insignificance of aught else. He obviously perceives the bearing of his arguments against the alterability of the course of outward nature by prayer, on the question of miracles in general; for he is careful to say that " the possibility of miracles, of a rare and unusual transcendence of the world order is not here in question " (p. 38). It may be permitted me to suppose, however, that, if miracles were in question, the speaker who warns us " that we must look for the heart of the absolute religion in that part of it which prescribes our moral and religious relations " (p. 46) would not be disposed to advise those who had found the heart of Christianity to take much thought about its miraculous integument.

My anonymous sermon will have nothing to do with such notions as these, and its preacher is not too polite, to say nothing of charitable, towards those who entertain them.

Scientific men, therefore, are perfectly right in asserting that Christianity rests on miracles. If miracles never happened, Christianity, in any sense which is not a mockery, which does not make the term of none effect, has no reality. I dwell on this because there is now an effort making to get up a non-miraculous, invertebrate Christianity, which may escape the ban of science. And I would warn you very distinctly against this new contrivance. Christianity is essentially miraculous, and falls to the ground if miracles be impossible.

Well, warning for warning. I venture to warn this preacher and those who, with him, persist in identifying Christianity with the miraculous, that such forms of Christianity are not only doomed to fall to the ground; but that, within the last half century, they have been driving that way with continually accelerated velocity.

The so-called religious world is given to a strange delusion. It fondly imagines that it possesses the monopoly of serious and constant reflection upon the terrible problems of existence ; and that those who cannot accept its shibboleths are either mere Gallios, caring for none of these things, or libertines desiring to escape from the restraints of morality. It does not appear to have entered the imaginations of these people that, outside their pale and firmly resolved never to enter it, there are thousands of men, certainly not their inferiors in character, capacity, or knowledge of the questions at issue, who estimate those purely spiritual elements of the Christian faith of which the Bishop of Manchester speaks as highly as the Bishop does ; but who will have nothing to do with the Christian Churches, because in their apprehension and for them, the profession of belief in the miraculous, on the evidence offered, would be simply immoral.

So far as my experience goes, men of science are neither better nor worse than the rest of the world. Occupation with the endlessly great parts of the universe does not necessarily involve greatness of character, nor does microscopic study of the infinitely little always produce humility. We have our full share of original sin ; need, greed, and vainglory beset us as they do other mortals; and our progress is, for the most past, like that of a tacking ship, the resultant of opposite divergencies from the straight path. But, for all that, there is one moral benefit which the pursuit of science unquestionably bestows. It keeps the estimate of the value of evidence up to the proper mark; and we are constantly receiving lessons, and sometimes very sharp ones, on the nature of proof. Men of science will always act up to their standard of veracity, when mankind in general leave off sinning; but that standard appears to me to be higher among them than in any other class of the community.

I do not know any body of scientific men who could be got to listen without the strongest expressions of disgusted repudiation to the exposition of a pretended scientific discovery, which had no better evidence to show for itself than the story of the devils entering a herd of swine, or of the fig-tree that was blasted for bearing no figs when "it was not the season of figs." Whether such events are possible or impossible, no man can say; but scientific ethics can and does declare that the profession of belief in them, on the evidence of documents of unknown date and of unknown authorship, is immoral. Theological apologists who insist that morality will vanish if their dogmas are exploded, would do well to consider the fact that, in the matter of intellectual veracity, science is already a long way ahead of the

Churches; and that, in this particular, it is exerting an educational influence on mankind of which the Churches have shown themselves utterly incapable.

Undoubtedly that varying compound of some of the best and some of the worst elements of Paganism and Judaism, moulded in practice by the innate character of certain people of the Western world, which, since the second century, has assumed to itself the title of orthodox Christianity, "rests on miracles" and falls to the ground, not "if miracles be impossible," but if those to which it is committed prove themselves unable to fulfil the conditions of honest belief. That this Christianity is doomed to fall is, to my mind, beyond a doubt; but its fall will be neither sudden nor speedy. The Church, with all the aid lent it by the secular arm, took many centuries to extirpate the open practice of pagan idolatry within its own fold; and those who have travelled in southern Europe will be aware that it has not extirpated the essence of such idolatry even yet. *Mutato nomine*, it is probable that there is as much sheer fetichism among the Roman populace now as there was eighteen hundred years ago; and if Marcus Antoninus could descend from his horse and ascend the steps of the Ara Cœli church about Twelfth Day, the only thing that need strike him would be the extremely contemptible character of the modern idols as works of art.

Science will certainly neither ask for, nor receive, the aid of the secular arm. It will trust to the much better and more powerful help of that education in scientific truth and in the morals of assent, which is rendered as indispensable, as it is inevitable, by the permeation of practical life with the products and ideas of science. But no one who considers the present state of even the most developed countries can doubt that the scientific light that has come into the world will have to shine in the midst of darkness for a long time. The urban populations, driven into contact with science by trade and manufacture, will more and more receive it, while the *pagani* will lag behind. Let us hope that no Julian may arise among them to head a forlorn hope against the inevitable. Whatever happens, science may bide her time in patience and in confidence.

But to return to my "Anonymous." I am afraid that if he represents any great party in the Church, the spirit of justice and reasonableness which animates the three bishops has as slender a chance of being imitated, on a large scale, as their common sense and their courtesy. For, not contented with misrepresenting science on its speculative side, "Anonymous" attacks its morality.

For two whole years, investigations and conclusions which would upset the theories of Darwin on the formation of coral islands were actually suppressed, and that by the advice even of those who accepted them, *for fear of upsetting the faith and dis*

*turbing the judgment formed by the multitude on the scientific
character—the infallibility—of the great master!*

So far as I know anything about the matters
which are here referred to, the part of this passage
which I have italicised is absolutely untrue. I
believe that I am intimately acquainted with all
Mr. Darwin's immediate scientific friends: and I
say that no one of them, nor any other man of
science known to me, ever could, or would, have
given such advice to any one—if for no other
reason than that, with the example of the most
candid and patient listener to objections that ever
lived fresh in their memories, they could not so
grossly have at once violated their highest duty
and dishonoured their friend.

The charge thus brought by "Anonymous"
affects the honour and the probity of men of
science; if it is true, we have forfeited all claim
to the confidence of the general public. In
my belief it is utterly false, and its real effect will
be to discredit those who are responsible for it.
As is the way with slanders, it has grown by
repetition. "Anonymous" is responsible for the
peculiarly offensive form which it has taken in his
hands; but he is not responsible for originating
it. He has evidently been inspired by an article
entitled "A Great Lesson," published in the September number of this Review. Truly it is "a
great lesson," but not quite in the sense intended
by the giver thereof.

In the course of his doubtless well-meant admonitions, the Duke of Argyll commits himself to a greater number of statements which are demonstrably incorrect and which any one who ventured to write upon the subject ought to have known to be incorrect, than I have ever seen gathered together in so small a space.

I submit a gathering from the rich store for the appreciation of the public.

First :—

Mr. Murray's new explanation of the structure of coral-reefs and islands was communicated to the Royal Society of Edinburgh in 1880, and supported with such a weight of facts and such a close texture of reasoning, that no serious reply has ever been attempted (p. 305).

"No serious reply has ever been attempted"! I suppose that the Duke of Argyll may have heard of Professor Dana, whose years of labour devoted to corals and coral-reefs when he was naturalist of the American expedition under Commodore Wilkes, more than forty years ago, have ever since caused him to be recognised as an authority of the first rank on such subjects. Now does his Grace know, or does he not know, that, in the year 1885, Professor Dana published an elaborate paper " On the Origin of Coral-Reefs and Islands," in which, after referring to a Presidential Address by the Director of the Geological Survey of Great Britain and Ireland delivered in 1883, in which special attention is directed to Mr. Murray's views Professor Dana says :—

The existing state of doubt on the question has led the writer to reconsider the earlier and later facts, and in the following pages he gives his results.

Professor Dana then devotes many pages of his very "serious reply" to a most admirable and weighty criticism of the objections which have at various times been raised to Mr. Darwin's doctrine, by Professor Semper, by Dr. Rein, and finally by Mr. Murray, and he states his final judgment as follows :—

With the theory of abrasion and solution incompetent, all the hypotheses of objectors to Darwin's theory are alike weak ; for all have made these processes their chief reliance, whether appealing to a calcareous, or a volcanic, or a mountain-peak basement for the structure. The subsidence which the Darwinian theory requires has not been opposed by the mention of any fact at variance with it, nor by setting aside Darwin's arguments in its favour ; and it has found new support in the facts from the "Challenger's" soundings off Tahiti, that had been put in array against it, and strong corroboration in the facts from the West Indies.

Darwin's theory, therefore, remains as the theory that accounts for the origin of reefs and islands.[1]

Be it understood that I express no opinion on the controverted points. I doubt if there are ten living men who, having a practical knowledge of what a coral-reef is, have endeavoured to master the very difficult biological and geological problems involved in their study. I happen to have

[1] *American Journal of Science*, 1885, p. 190.

spent the best part of three years among coral-
reefs and to have made that attempt; and, when
Mr. Murray's work appeared, I said to myself that
until I had two or three months to give to the
renewed study of the subject in all its bearings, I
must be content to remain in a condition of suspended judgment. In the meanwhile, the man
who would be voted by common acciamation as the
most competent person now living to act as umpire,
has delivered the verdict I have quoted; and, to
go no further, has fully justified the hesitation I
and others may have felt about expressing an
opinion. Under these circumstances, it seems to
me to require a good deal of courage to say "no
serious reply has ever been attempted"; and to
chide the men of science, in lofty tones, for their
"reluctance to admit an error" which is not
admitted; and for their "slow and sulky acquiescence" in a conclusion which they have the
gravest warranty for suspecting.

Second :—

Darwin himself had lived to hear of the new solution, and,
with that splendid candour which was eminent in him, his mind,
though now grown old in his own early convictions, was at least
ready to entertain it, and to confess that serious doubts had been
awakened as to the truth of his famous theory (p. 305).

I wish that Darwin's splendid candour could
be conveyed by some description of spiritual
" microbe" to those who write about him. I am
not aware that Mr. Darwin ever entertained

"serious doubts as to the truth of his famous theory"; and there is tolerably good evidence to the contrary. The second edition of his work, published in 1876, proves that he entertained no such doubts then; a letter to Professor Semper, whose objections, in some respects, forestalled those of Mr. Murray, dated October 2, 1879, expresses his continued adherence to the opinion "that the atolls and barrier reefs in the middle of the Pacific and Indian Oceans indicate subsidence"; and the letter of my friend Professor Judd, printed at the end of this article (which I had perhaps better say Professor Judd had not seen) will prove that this opinion remained unaltered to the end of his life.

Third :—

. . . Darwin's theory is a dream. It is not only unsound, but it is in many respects the reverse of truth. With all his conscientiousness, with all his caution, with all his powers of observation, Darwin in this matter fell into errors as profound as the abysses of the Pacific (p. 301).

Really ? It seems to me that, under the circumstances, it is pretty clear that these lines exhibit a lack of the qualities justly ascribed to Mr. Darwin, which plunges their author into a much deeper abyss, and one from which there is no hope of emergence.

Fourth :—

All the acclamations with which it was received were as the shouts of an ignorant mob (p. 301)

But surely it should be added that the Coryphæus
of this ignorant mob, the fugleman of the shouts,
was one of the most accomplished naturalists and
geologists now living—the American Dana—who,
after years of independent study extending over
numerous reefs in the Pacific, gave his hearty
assent to Darwin's views, and after all that had
been said, deliberately reaffirmed that assent in
the year 1885.

Fifth :—

> The overthrow of Darwin's speculation is only beginning to
> be known. It has been whispered for some time. The cherished
> dogma has been dropping very slowly out of sight (p. 301).

Darwin's speculation may be right or wrong, but I
submit that that which has not happened cannot
even begin to be known, except by those who have
miraculous gifts to which we poor scientific people
do not aspire.   The overthrow of Darwin's views
may have been whispered by those who hoped for
it; and they were perhaps wise in not raising
their voices above a whisper.   Incorrect statements, if made too loudly, are apt to bring about
unpleasant consequences.

Sixth :—

Mr. Murray's views, published in 1880, are
said to have met with "slow and sulky acquiescence " (p. 305).   I have proved that they
cannot be said to have met with general acquiescence of any sort, whether quick and cheerful,
or slow and sulky ; and if this assertion is meant

to convey the impression that Mr. Murray's views have been ignored, that there has been a conspiracy of silence against them, it is utterly contrary to notorious fact.

Professor Geikie's well-known "Textbook of Geology" was published in 1882, and at pages 457-459 of that work there is a careful exposition of Mr. Murray's views. Moreover Professor Geikie has specially advocated them on other occasions,[1] notably in a long article on "The Origin of Coral-Reefs," published in two numbers of "Nature" for 1883, and in a Presidential Address delivered in the same year. If, in so short a time after the publication of his views, Mr. Murray could boast of a convert, so distinguished and influential as the Director of the Geological Survey, it seems to me that this wonderful *conspiration de silence* (which has about as much real existence as the Duke of Argyll's other bogie, "The Reign of Terror") must have *ipso facto* collapsed. I wish that, when I was a young man, my endeavours to upset some prevalent errors had met with as speedy and effectual backing.

Seventh :—

. . . Mr. John Murray was strongly advised against the publication of his views in derogation of Darwin's long-accepted

---

[1] Professor Geikie, however, though a strong, is a fair and candid advocate. He says of Darwin's theory, "That it may be possibly true, in some instances, may be readily granted." For Professor Geikie, then, it is not yet overthrown—still less a dream.

theory of the coral islands, and was actually induced to delay it for two years. Yet the late Sir Wyville Thomson, who was at the head of the naturalists of the " Challenger " expedition, was himself convinced by Mr. Murray's reasoning (p. 307).

Clearly, then, it could not be Mr. Murray's official chief who gave him this advice. Who was it? And what was the exact nature of the advice given? Until we have some precise information on this head, I shall take leave to doubt whether this statement is more accurate than those which I have previously cited.

Whether such advice was wise or foolish, just or immoral, depends entirely on the motive of the person who gave it. If he meant to suggest to Mr. Murray that it might be wise for a young and comparatively unknown man to walk warily, when he proposed to attack a generalisation based on many years' labour of one undoubtedly competent person, and fortified by the independent results of the many years' labour of another undoubtedly competent person ; and even, if necessary, to take two whole years in fortifying his position, I think that such advice would have been sagacious and kind. I suppose that there are few working men of science who have not kept their ideas to themselves, while gathering and sifting evidence, for a much longer period than two years.

If, on the other hand, Mr. Murray was advised to delay the publication of his criticisms, simply to

save Mr. Darwin's credit and to preserve some
reputation for infallibility, which no one ever
heard of, then I have no hesitation in declaring
that his adviser was profoundly dishonest, as well
as extremely foolish; and that, if he is a man of
science, he has disgraced his calling.

But, after all, this supposed scientific Achitophel
has not yet made good the primary fact of his existence. Until the needful proof is forthcoming, I
think I am justified in suspending my judgment as
to whether he is much more than an anti-scientific
myth. I leave it to the Duke of Argyll to judge
of the extent of the obligation under which, for
his own sake, he may lie to produce the evidence
on which his aspersions of the honour of scientific
men are based. I cannot pretend that we are
seriously disturbed by charges which every one
who is acquainted with the truth of the matter
knows to be ridiculous; but mud has a habit of
staining if it lies too long, and it is as well to have
it brushed off as soon as may be.

So much for the "Great Lesson." It is followed
by a "Little Lesson," apparently directed against
my infallibility—a doctrine about which I should
be inclined to paraphrase Wilkes's remark to
George the Third, when he declared that he, at
any rate, was not a Wilkite. But I really should
be glad to think that there are people who need
the warning, because then it will be obvious that
this raking up of an old story cannot have been

suggested by a mere fanatical desire to damage men of science. I can but rejoice, then, that these misguided enthusiasts, whose faith in me has so far exceeded the bounds of reason, should be set right. But that "want of finish" in the matter of accuracy which so terribly mars the effect of the "Great Lesson," is no less conspicuous in the case of the "Little Lesson," and, instead of setting my too fervent disciples right, it will set them wrong.

The Duke of Argyll, in telling the story of *Bathybius*, says that my mind was "caught by this new and grand generalisation of the physical basis of life." I never have been guilty of a reclamation about anything to my credit, and I do not mean to be; but if there is any blame going, I do not choose to be relegated to a subordinate place when I have a claim to the first. The responsibility for the first description and the naming of *Bathybius* is mine and mine only. The paper on "Some Organisms living at great Depths in the Atlantic Ocean," in which I drew attention to this substance, is to be found by the curious in the eighth volume of the "Quarterly Journal of Microscopical Science," and was published in the year 1868. Whatever errors are contained in that paper are my own peculiar property; but neither at the meeting of the British Association in 1868, nor anywhere else, have I gone beyond what is there stated; except in so far that, at a long-sub-

sequent meeting of the Association, being importuned about the subject, I ventured to express
somewhat emphatically, the wish that the thing
was at the bottom of the sea.

What is meant by my being caught by a
generalisation about the physical basis of life I
do not know; still less can I understand the assertion that *Bathybius* was accepted because of its
supposed harmony with Darwin's speculations.
That which interested me in the matter was the
apparent analogy of *Bathybius* with other well-
known forms of lower life, such as the plasmodia
of the Myxomycetes and the Rhizopods. Speculative hopes or fears had nothing to do with the
matter; and if *Bathybius* were brought up alive
from the bottom of the Atlantic to-morrow, the
fact would not have the slightest bearing, that I
can discern, upon Mr. Darwin's speculations, or
upon any of the disputed problems of biology. It
would merely be one elementary organism the
more added to the thousands already known.

Up to this moment I was not aware of the
universal favour with which *Bathybius* was received.[1]   Those simulators of an "ignorant mob"
who, according to the Duke of Argyll, welcomed

----

[1] I find, moreover, that I specially warned my readers against
hasty judgment. After stating the facts of observation, I add,
"I have, hitherto, said nothing about their meaning, as, in an
inquiry so difficult and fraught with interest as this, it seems to
me to be in the highest degree important to keep the questions
of fact and the questions of interpretation well apart" (p. 210).

Darwin's theory of coral-reefs, made no demonstration in my favour, unless his Grace includes Sir Wyville Thomson, Dr. Carpenter, Dr. Bessels, and Professor Haeckel under that head. On the contrary, a sagacious friend of mine, than whom there was no more competent judge, the late Mr. George Busk, was not to be converted; while, long before the "Challenger" work, Ehrenberg wrote to me very sceptically; and I fully expected that that eminent man would favour me with pretty sharp criticism. Unfortunately, he died shortly afterwards, and nothing from him, that I know of, appeared. When Sir Wyville Thomson wrote to me a brief account of the results obtained on board the "Challenger" I sent this statement to "Nature," in which journal it appeared the following week, without any further note or comment than was needful to explain the circumstances. In thus allowing judgment to go by default, I am afraid I showed a reckless and ungracious disregard for the feelings of the believers in my infallibility. No doubt I ought to have hedged and fenced and attenuated the effect of Sir Wyville Thomson's brief note in every possible way. Or perhaps I ought to have suppressed the note altogether, on the ground that it was a mere *ex parte* statement. My excuse is that, notwithstanding a large and abiding faith in human folly, I did not know then, any more than I know now, that there was anybody foolish enough to be unaware that

has dropped into the abyss when it is quite obviously alive and kicking at the surface; he must not assimilate a man like Professor Dana to the components of an "ignorant mob"; he must not say that things are beginning to be known which are not known at all; he must not say that "slow and sulky acquiescence" has been given to that which cannot yet boast of general acquiescence of any kind; he must not suggest that a view which has been publicly advocated by the Director of the Geological Survey and no less publicly discussed by many other authoritative writers has been intentionally and systematically ignored; he must not ascribe ill motives for a course of action which is the only proper one; and finally, if any one but myself were interested, I should say that he had better not waste his time in raking up the errors of those whose lives have been occupied, not in talking about science, but in toiling, sometimes with success and sometimes with failure, to get some real work done.

The most considerable difference I note among men is not in their readiness to fall into error, but in their readiness to acknowledge these inevitable lapses. The Duke of Argyll has now a splendid opportunity for proving to the world in which of these categories it is hereafter to rank him.

DEAR PROFESSOR HUXLEY,—A short time before Mr. Darwin's death. I had a conversation with him concerning the observations which had been made by Mr. Murray upon coral-reefs, and the speculations which had been founded upon those observations. I found that Mr. Darwin had very carefully considered the whole subject, and that while, on the one hand, he did not regard the actual facts recorded by Mr. Murray as absolutely inconsistent with his own theory of subsidence, on the other hand, he did not believe that they necessitated or supported the hypothesis advanced by Mr. Murray. Mr. Darwin's attitude, as I understood it, towards Mr. Murray's objections to the theory of subsidence was exactly similar to that maintained by him with respect to Professor Semper's criticism, which was of a very similar character; and his position with regard to the whole question was almost identical with that subsequently so clearly defined by Professor Dana in his well-known articles published in the "American Journal of Science" for 1885.

It is difficult to imagine how any one, acquainted with the scientific literature of the last seven years, could possibly suggest that Mr. Murray's memoir published in 1880 had failed to secure a due amount of attention. Mr. Murray, by his position in the "Challenger" office, occupied an exceptionally favourable position for making his views widely known; and he had, moreover, the singular good fortune to secure from the first the advocacy of so able and brilliant a writer as

Professor Archibald Geikie, who in a special discourse and in several treatises on geology and
physical geology very strongly supported the new
theory.   It would be an endless task to attempt
to give references to the various scientific journals
which have discussed the subject, but I may add
that every treatise on geology which has been
published, since Mr. Murray's views were made
known, has dealt with his observations at considerable length.   This is true of Professor A. H.
Green's "Physical Geology," published in 1882;
of Professor Prestwich's "Geology, Chemical and
Physical"; and of Professor James Geikie's "Out-
lines of Geology," published in 1886.   Similar
prominence is given to the subject in De Lap-
parent's "Traité de Geologie," published in 1885,
and in Credner's "Elemente der Geologie," which
has appeared during the present year.   If this be
a "conspiracy of silence," where, alas! can the
geological speculator seek for fame?—Yours very
truly,   JOHN W. JUDD.

October 10, 1887.

V

THE VALUE OF WITNESS TO THE MIRACULOUS

[1889]

CHARLES, or, more properly, Karl, King of the Franks, consecrated Roman Emperor in St. Peter's on Christmas Day, A.D. 800, and known to posterity as the Great (chiefly by his agglutinative Gallicised denomination of Charlemagne), was a man great in all ways, physically and mentally. Within a couple of centuries after his death Charlemagne became the centre of innumerable legends ; and the myth-making process does not seem to have been sensibly interfered with by the existence of sober and truthful histories of the Emperor and of the times which immediately preceded and followed his reign, by a contemporary writer who occupied a high and confidential position in his court, and in that of his successor. This was one Eginhard, or Einhard, who appears to have been born about A.D. 770, and spent his youth at the court, being educated along with Charles's sons. There is excellent contemporary testimony not only to Eginhard's existence, but to his abilities, and to the place which he occupied in the circle of the intimate friends of the great ruler whose life he subsequently wrote. In fact, there is as good evidence of Eginhard's existence, of his official position, and of his being the author of the chief works attributed to him, as can reasonably be expected in the case of a man who lived more than a thousand years ago, and was neither a great king nor a great warrior. The works are—1. "The Life of the Emperor Karl." 2. "The Annals of the Franks." 3. "Letters." 4. "The History of the Translation of the Blessed Martyrs of Christ, SS. Marcellinus and Petrus."

It is to the last, as one of the most singular and interesting records of the period during which the Roman world passed into that of the Middle Ages, that I wish to direct attention.[1] It was written in the ninth century, somewhere, apparently, about the year 830, when Eginhard, ailing in health and weary of political life, had withdrawn to the monastery of Seligenstadt, of which he was the founder. A manuscript copy of the work, made in the tenth century, and once the

---

[1] My citations are made from Teulet's *Einhardi omnia quæ extant opera*, Paris, 1840-1843, which contains a biography of the author, a history of the text, with translations into French, and many valuable annotations.

property of the monastery of St. Bavon on the
Scheldt, of which Eginhard was Abbot, is still
extant, and there is no reason to believe that, in
this copy, the original has been in any way interpolated or otherwise tampered with. The main
features of the strange story contained in the
"Historia Translationis" are set forth in the
following pages, in which, in regard to all matters
of importance, I shall adhere as closely as possible
to Eginhard's own words.

While I was still at Court, busied with secular affairs, I often
thought of the leisure which I hoped one day to enjoy in a
solitary place, far away from the crowd, with which the liber-
ality of Prince Louis, whom I then served, had provided me.
This place is situated in that part of Germany which lies between
the Neckar and the Maine,[1] and is nowadays called the Odenwald by those who live in and about it. And here having built,
according to my capacity and resources, not only houses and
permanent dwellings, but also a basilica fitted for the performance of divine service and of no mean style of construction, I
began to think to what saint or martyr I could best dedicate it.
A good deal of time had passed while my thoughts fluctuated
about this matter, when it happened that a certain deacon of
the Roman Church, named Deusdona, arrived at the Court for
the purpose of seeking the favour of the King in some affairs in
which he was interested. He remained some time ; and then,
having transacted his business, he was about to return to Rome
when one day, moved by courtesy to a stranger, we invited
him to a modest refection ; and while talking of many things
at table, mention was made of the translation of the body of
the blessed Sebastian, and of the neglected tombs of the

---

[1] At present included in the Duchies of Hesse-Darmstadt and
Baden.
[2] This took place in the year 826 A.D.   The relics were
brought from Rome and deposited in the Church of St. Medardus
at Soissons.

martyrs, of which there is such a prodigious number at Rome ;
and the conversation having turned towards the dedication of
our new basilica, I began to inquire how it might be possible for
me to obtain some of the true relics of the saints which rest at
Rome. He at first hesitated, and declared that he did not know
how that could be done. But observing that I was both anxious
and curious about the subject, he promised to give me an answer
some other day.

When I returned to the question some time afterwards, he immediately drew from his bosom a paper, which he begged me to
read when I was alone, and to tell him what I was disposed to
think of that which was therein stated. I took the paper and,
as he desired, read it alone and in secret. (Cap. i. 2, 3.)

I shall have occasion to return to Deacon
Deusdona's conditions, and to what happened
after Eginhard's acceptance of them. Suffice it,
for the present, to say that Eginhard's notary,
Ratleicus (Ratleig), was despatched to Rome and
succeeded in securing two bodies, supposed to be
those of the holy martyrs Marcellinus and Petrus ;
and when he had got as far on his homeward
journey as the Burgundian town of Solothurn,
or Soleure,[1] notary Ratleig despatched to his
master, at St. Bavon, a letter announcing the
success of his mission.

As soon as by reading it I was assured of the arrival of the
saints, I despatched a confidential messenger to Maestricht to
gather together priests, other clerics, and also laymen, to go out
to meet the coming saints as speedily as possible. And he and
his companions, having lost no time, after a few days met those
who had charge of the saints at Solothurn. Joined with them,

---

[1] Now included in Western Switzerland.

and with a vast crowd of people who gathered from all parts, singing hymns, and amidst great and universal rejoicings, they travelled quickly to the city of Argentoratum, which is now called Strasburg.   Thence embarking on the Rhine, they came to the place called Portus,[1] and landing on the east bank of the river, at the fifth station thence they arrived at Michilinstadt,[2] accompanied by an immense multitude, praising God.   This place is in that forest of Germany which in modern times is called the Odenwald, and about six leagues from the Maine.   And here, having found a basilica recently built by me, but not yet consecrated, they carried the sacred remains into it and deposited them therein, as if it were to be their final resting-place.   As soon as all this was reported to me I travelled thither as quickly as I could.   (Cap. ii. 14.)

Three days after Eginhard's arrival began the series of wonderful events which he narrates, and for which we have his personal guarantee.   The first thing that he notices is the dream of a servant of Ratleig, the notary, who, being set to watch the holy relics in the church after vespers, went to sleep and, during his slumbers, had a vision of two pigeons, one white and one gray and white, which came and sat upon the bier over the relics ; while, at the same time, a voice ordered the man to tell his master that the holy martyrs had chosen another resting-place and desired to be transported thither without delay.

Unfortunately, the saints seem to have forgotten to mention where they wished to go ; and, with the most anxious desire to gratify their

---

[1] Probably, according to Teulet, the present Sandhofer-fahrt, a little below the embouchure of the Neckar.

[2] The present Michilstadt, thirty miles N.E. of Heidelberg.

smallest wishes, Eginhard was naturally greatly
perplexed what to do. While in this state of
mind, he was one day contemplating his "great
and wonderful treasure, more precious than all
the gold in the world," when it struck him that
the chest in which the relics were contained was
quite unworthy of its contents ; and, after vespers,
he gave orders to one of the sacristans to take the
measure of the chest in order that a more fitting
shrine might be constructed. The man, having
lighted a wax candle and raised the pall which
covered the relics, in order to carry out his
master's orders, was astonished and terrified to
observe that the chest was covered with a blood-
like exudation (*loculum mirum in modum humore
sanguineo undique distillantem*), and at once sent
a message to Eginhard.

> Then I and those priests who accompanied me beheld this
> stupendous miracle, worthy of all admiration. For just as when
> it is going to rain, pillars and slabs and marble images exude
> moisture, and, as it were, sweat, so the chest which contained
> the most sacred relics was found moist with the blood exuding
> on all sides. (Cap. ii. 16.)

Three days' fast was ordained in order that the
meaning of the portent might be ascertained. All
that happened, however, was that, at the end of
that time, the "blood," which had been exuding in
drops all the while, dried up. Eginhard is careful
to say that the liquid " had a saline taste, something like that of tears, and was thin as water,

though of the colour of true blood," and he clearly
thinks this satisfactory evidence that it was
blood.

The same night, another servant had a vision, in
which still more imperative orders for the removal
of the relics were given ; and, from that time forth,
" not a single night passed without one, two, or
even three of our companions receiving revelations
in dreams that the bodies of the saints were to be
transferred from that place to another." At last a
priest, Hildfrid, saw, in a dream, a venerable
white-haired man in a priest's vestments, who
bitterly reproached Eginhard for not obeying the
repeated orders of the saints ; and, upon this, the
journey was commenced. Why Eginhard delayed
obedience to these repeated visions so long does
not appear. He does not say so, in so many words,
but the general tenor of the narrative leads one to
suppose that Mulinheim (afterwards Seligenstadt)
is the "solitary place" in which he had built the
church which awaited dedication. In that case,
all the people about him would know that he
desired that the saints should go there. If a
glimmering of secular sense led him to be a little
suspicious about the real cause of the unanimity of
the visionary beings who manifested themselves to
his *entourage* in favour of moving on, he does not
say so.

At the end of the first day's journey, the precious
relics were deposited in the church of St. Martin,

in the village of Ostheim. Hither, a paralytic nun
(*sanctimonialis quædam paralytica*) of the name of
Ruodlang was brought, in a car, by her friends and
relatives from a monastery a league off. She spent
the night watching and praying by the bier of the
saints; " and health returning to all her members,
on the morrow she went back to her place whence
she came, on her feet, nobody supporting her, or
in any way giving her assistance." (Cap. ii. 19.)

On the second day, the relics were carried to
Upper Mulinheim; and, finally, in accordance with
the orders of the martyrs, deposited in the church
of that place, which was therefore renamed
Seligenstadt. Here, Daniel, a beggar boy of fifteen,
and so bent that " he could not look at the sky
without lying on his back," collapsed and fell down
during the celebration of the Mass. " Thus he lay
a long time, as if asleep, and all his limbs straight-
ening and his flesh strengthening (*recepta firmitate
nervorum*), he arose before our eyes, quite well."
(Cap. ii. 20.)

Some time afterwards an old man entered the
church on his hands and knees, being unable to
use his limbs properly :—

He, in presence of all of us, by the power of God and the
merits of the blessed martyrs, in the same hour in which he
entered was so perfectly cured that he walked without so much
as a stick. And he said that, though he had been deaf for five
years, his deafness had ceased along with the palsy. (Cap. iii.
33.)

Eginhard was now obliged to return to the
Court at Aix-la-Chapelle, where his duties kept
him through the winter; and he is careful to point
out that the later miracles which he proceeds to
speak of are known to him only at second hand.
But, as he naturally observes, having seen such
wonderful events with his own eyes, why should
he doubt similar narrations when they are received from trustworthy sources?

Wonderful stories these are indeed, but as they
are, for the most part, of the same general character
as those already recounted, they may be passed
over.   There is, however, an account of a possessed
maiden which is worth attention.   This is set forth
in a memoir, the principal contents of which are
the speeches of a demon who declared himself to
possess the singular appellation of " Wiggo," and
revealed himself in the presence of many witnesses,
before the altar, close to the relics of the blessed
martyrs.   It is noteworthy that the revelations
appear to have been made in the shape of replies
to the questions of the exorcising priest; and there
is no means of judging how far the answers are,
really, only the questions to which the patient replied yes or no.

The possessed girl, about sixteen years of age,
was brought by her parents to the basilica of the
martyrs.

When she approached the tomb containing the sacred bodies,
the priest, according to custom, read the formula of exorcism

over her head. When he began to ask how and when the
demon had entered her, she answered, not in the tongue of
the barbarians, which alone the girl knew, but in the Roman
tongue. And when the priest was astonished and asked how
she came to know Latin, when her parents, who stood by, were
wholly ignorant of it, "Thou hast never seen my parents," was
the reply. To this the priest, "Whence art thou, then, if these
are not thy parents ?" And the demon, by the mouth of the
girl, "I am a follower and disciple of Satan, and for a long time
I was gatekeeper (janitor) in hell ; but, for some years, along
with eleven companions, I have ravaged the kingdom of the
Franks." (Cap. v. 49.)

He then goes on to tell how they blasted the
crops and scattered pestilence among beasts and
men, because of the prevalent wickedness of the
people.[1]

The enumeration of all these iniquities, in
oratorical style, takes up a whole octavo page ; and
at the end it is stated, "All these things the
demon spoke in Latin by the mouth of the girl."

And when the priest imperatively ordered him to come out,
"I shall go," said he, "not in obedience to you, but on account
of the power of the saints, who do not allow me to remain any
longer." And, having said this, he threw the girl down on the
floor and there compelled her to lie prostrate for a time, as
though she slumbered. After a little while, however, he going
away, the girl, by the power of Christ and the merits of the
blessed martyrs, as it were awaking from sleep, rose up quite
well, to the astonishment of all present ; nor after the demon
had gone out was she able to speak Latin : so that it was plain
enough that it was not she who had spoken in that tongue, but
the demon by her mouth. (Cap. v. 51.)

---

[1] In the Middle Ages one of the most favourite accusations
against witches was that they committed just these enormities.

If the "Historia Translationis" contained nothing
more than has been laid before the reader, up to this
time, disbelief in the miracles of which it gives
so precise and full a record might well be regarded
as hyper-scepticism. It might fairly be said, Here
you have a man, whose high character, acute intelligence, and large instruction are certified by
eminent contemporaries ; a man who stood high in
the confidence of one of the greatest rulers of any
age, and whose other works prove him to be an
accurate and judicious narrator of ordinary events.
This man tells you, in language which bears the
stamp of sincerity, of things which happened within
his own knowledge, or within that of persons in
whose veracity he has entire confidence, while he
appeals to his sovereign and the court as witnesses
of others ; what possible ground can there be for
disbelieving him ?

Well, it is hard upon Eginhard to say so, but it
is exactly the honesty and sincerity of the man
which are his undoing as a witness to the miraculous. He himself makes it quite obvious that
when his profound piety comes on the stage, his
good sense and even his perception of right and
wrong, make their exit. Let us go back to the
point at which we left him, secretly perusing the
letter of Deacon Deusdona. As he tells us, its
contents were

that he [the deacon] had many relics of saints at home, and that
he would give them to me if I would furnish him with the

means of returning to Rome; he had observed that 1 had two
mules, and if 1 would let him have one of them and would
despatch with him a confidential servant to take charge of the
relics, he would at once send them to me.   This plausibly expressed proposition pleased me, and I made up my mind to test
the value of the somewhat ambiguous promise at once; [1] so
giving him the mule and money for his journey I ordered my
notary Ratleig (who already desired to go to Rome to offer his
devotions there) to go with him.   Therefore, having left Aix-
la-Chapelle (where the Emperor and his Court resided at the
time) they came to Soissons.   Here they spoke with Hildoin,
abbot of the monastery of St. Medardus, because the said deacon
had assured him that he had the means of placing in his possession the body of the blessed Tiburtius the Martyr.   Attracted
by which promises he (Hildoin) sent with them a certain priest,
Hunus by name, a sharp man (*hominem callidum*), whom he
ordered to receive and bring back the body of the martyr in
question.   And so, resuming their journey, they proceeded to
Rome as fast as they could.   (Cap. i. 3.)

Unfortunately, a servant of the notary, one
Reginbald, fell ill of a tertian fever, and impeded
the progress of the party.   However, this piece of
adversity had its sweet uses; for three days before
they reached Rome, Reginbald had a vision.
Somebody habited as a deacon appeared to him
and asked why his master was in such a hurry to
get to Rome; and when Reginbald explained their
business, this visionary deacon, who seems to have
taken the measure of his brother in the flesh with
some accuracy, told him not by any means to

---

[1] It is pretty clear that Eginhard had his doubts about the
deacon, whose pledges he qualifies as *sponsiones incertæ*.   But,
to be sure, he wrote after events which fully justified scepticism.

expect that Deusdona would fulfil his promises.
Moreover, taking the servant by the hand, he led
him to the top of a high mountain and, showing
him Rome (where the man had never been),
pointed out a church, adding " Tell Ratleig the
thing he wants is hidden there ; let him get it as
quickly as he can and go back to his master."
By way of a sign that the order was authoritative, the servant was promised that, from that
time forth, his fever should disappear.   And as
the fever did vanish to return no more, the faith
of Eginhard's people in Deacon Deusdona naturally vanished with it (*et fidem diaconi promissis
non haberent*).   Nevertheless, they put up at the
deacon's house near St. Peter ad Vincula.   But
time went on and no relics made their appearance,
while the notary and the priest were put off with
all sorts of excuses—the brother to whom the
relics had been confided was gone to Beneventum
and not expected back for some time, and so on
—until Ratleig and Hunus began to despair, and
were minded to return, *infecto negotio*.

But my notary, calling to mind his servant's dream, proposed
to his companion that they should go to the cemetery which
their host had talked about without him.   So, having found and
hired a guide, they went in the first place to the basilica of the
blessed Tiburtius in the Via Labicana, about three thousand
paces from the town, and cautiously and carefully inspected the
tomb of that martyr, in order to discover whether it could be
opened without any one being the wiser.   Then they descended
into the adjoining crypt, in which the bodies of the blessed

martyrs of Christ, Marcellinus and Petrus, were buried ; and,
having made out the nature of their tomb, they went away
thinking their host would not know what they had been about.
But things fell out differently from what they had imagined.
(Cap. i. 7.)

In fact, Deacon Deusdona, who doubtless kept
an eye on his guests, knew all about their
manœuvres and made haste to offer his services, in
order that, " with the help of God " (*si Deus votis
eorum favere dignaretur*), they should all work
together.   The deacon was evidently alarmed lest
they should succeed without *his* help.

So, by way of preparation for the contemplated *vol avec effraction* they fasted three days;
and then, at night, without being seen, they betook themselves to the basilica of St. Tiburtius,
and tried to break open the altar erected over
his remains.   But the marble proving too solid,
they descended to the crypt, and, " having evoked
our Lord Jesus Christ and adored the holy
martyrs," they proceeded to prise off the stone
which covered the tomb, and thereby exposed the
body of the most sacred martyr, Marcellinus,
" whose head rested on a marble tablet on which
his name was inscribed."   The body was taken
up with the greatest veneration, wrapped in a rich
covering, and given over to the keeping of the
deacon and his brother, Lunison, while the stone
was replaced with such care that no sign of the
theft remained.

As sacrilegious proceedings of this kind were
punishable with death by the Roman law, it
seems not unnatural that Deacon Deusdona should
have become uneasy, and have urged Ratleig to be
satisfied with what he had got and be off with his
spoils.   But the notary having thus cleverly
captured the blessed Marcellinus, thought it a
pity he should be parted from the blessed Petrus,
side by side with whom he had rested, for five
hundred years and more, in the same sepulchre (as
Eginhard pathetically observes); and the pious
man could neither eat, drink, nor sleep, until he
had compassed his desire to re-unite the saintly
colleagues.   This time, apparently in consequence
of Deusdona's opposition to any further resurrec-
tionist doings, he took counsel with a Greek monk,
one Basil, and, accompanied by Hunus, but saying
nothing to Deusdona, they committed another
sacrilegious burglary, securing this time, not only
the body of the blessed Petrus, but a quantity of
dust, which they agreed the priest should take,
and tell his employer that it was the remains of the
blessed Tiburtius.   How Deusdona was " squared,"
and what he got for his not very valuable complicity in these transactions, does not appear.   But
at last the relics were sent off in charge of Lunison,
the brother of Deusdona, and the priest Hunus, as
far as Pavia, while Ratleig stopped behind for a
week to see if the robbery was discovered, and,
presumably, to act as a blind, if any hue and cry

was raised. But, as everything remained quiet, the
notary betook himself to Pavia, where he found
Lunison and Hunus awaiting his arrival. The
notary's opinion of the character of his worthy
colleagues, however, may be gathered from the
fact that, having persuaded them to set out in
advance along a road which he told them he was
about to take, he immediately adopted another
route, and, travelling by way of St. Maurice and
the Lake of Geneva, eventually reached Soleure.

Eginhard tells all this story with the most naive
air of unconsciousness that there is anything
remarkable about an abbot, and a high officer of
state to boot, being an accessory, both before and
after the fact, to a most gross and scandalous act
of sacrilegious and burglarious robbery. And an
amusing sequel to the story proves that, where
relics were concerned, his friend Hildoin, another
high ecclesiastical dignitary, was even less scrupulous than himself.

On going to the palace early one morning, after
the saints were safely bestowed at Seligenstadt, he
found Hildoin waiting for an audience in the
Emperor's antechamber, and began to talk to him
about the miracle of the bloody exudation. In the
course of conversation, Eginhard happened to
allude to the remarkable fineness of the garment
of the blessed Marcellinus. Whereupon Abbot
Hildoin observed (to Eginhard's stupefaction) that
his observation was quite correct. Much astonished

at this remark from a person who was supposed
not to have seen the relics, Eginhard asked him
how he knew that? Upon this, Hildoin saw that
he had better make a clean breast of it, and he
told the following story, which he had received
from his priestly agent, Hunus. While Hunus and
Lunison were at Pavia, waiting for Eginhard's
notary, Hunus (according to his own account) had
robbed the robbers. The relics were placed in a
church; and a number of laymen and clerics, of
whom Hunus was one, undertook to keep watch
over them. One night, however, all the watchers,
save the wide-awake Hunus, went to sleep; and
then, according to the story which this "sharp"
ecclesiastic foisted upon his patron,

it was borne in upon his mind that there must be some great
reason why all the people, except himself, had suddenly become
somnolent; and, determining to avail himself of the opportunity
thus offered (*oblata occasione utendum*), he rose and, having
lighted a candle, silently approached the chests. Then, having
burnt through the threads of the seals with the flame of the
candle, he quickly opened the chests, which had no locks;[1] and,
taking out portions of each of the bodies which were thus exposed, he closed the chests and connected the burnt ends of the
threads with the seals again, so that they appeared not to have
been touched; and, no one having seen him, he returned to his
place. (Cap. iii. 23.)

Hildoin went on to tell Eginhard that Hunus at
first declared to him that these purloined relics

---

[1] The words are *scrinia sine clave*, which seems to mean
"having no key." But the circumstances forbid the idea of
breaking open.

belonged to St. Tiburtius ; but afterwards confessed, as a great secret, how he had come by
them, and he wound up his discourse thus :

They have a place of honour beside St. Medardus, where they
are worshipped with great veneration by all the people ; but
whether we may keep them or not is for your judgment. (Cap.
iii. 23.)

Poor Eginhard was thrown into a state of great
perturbation of mind by this revelation. An
acquaintance of his had recently told him of a
rumour that was spread about that Hunus had
contrived to abstract *all* the remains of SS.
Marcellinus and Petrus while Eginhard's agents
were in a drunken sleep; and that, while the real
relics were in Abbot Hildoin's hands at St.
Medardus, the shrine at Seligenstadt contained
nothing but a little dust. Though greatly annoyed
by this " execrable rumour, spread everywhere by
the subtlety of the devil," Eginhard had doubtless
comforted himself by his supposed knowledge of
its falsity, and he only now discovered how considerable a foundation there was for the scandal.
There was nothing for it but to insist upon the
return of the stolen treasures. One would have
thought that the holy man, who had admitted
himself to be knowingly a receiver of stolen goods,
would have made instant restitution and begged
only for absolution. But Eginhard intimates that
he had very great difficulty in getting his brother
abbot to see that even restitution was necessary.

Hildoin's proceedings were not of such a nature as to lead any one to place implicit confidence in anything he might say; still less had his agent, priest Hunus, established much claim to confidence; and it is not surprising that Eginhard should have lost no time in summoning his notary and Lunison to his presence, in order that he might hear what they had to say about the business. They, however, at once protested that priest Hunus's story was a parcel of lies, and that after the relics left Rome no one had any opportunity of meddling with them. Moreover, Lunison, throwing himself at Eginhard's feet, confessed with many tears what actually took place. It will be remembered that after the body of St. Marcellinus was abstracted from its tomb, Ratleig deposited it in the house of Deusdona, in charge of the latter's brother, Lunison. But Hunus, being very much disappointed that he could not get hold of the body of St. Tiburtius, and afraid to go back to his abbot empty-handed, bribed Lunison with four pieces of gold and five of silver to give him access to the chest. This Lunison did, and Hunus helped himself to as much as would fill a gallon measure (*vas sextarii mensuram*) of the sacred remains. Eginhard's indignation at the "rapine" of this "nequissimus nebulo" is exquisitely droll. It would appear that the adage about the receiver being as bad as the thief was not current in the ninth century.

Let us now briefly sum up the history of the acquisition of the relics. Eginhard makes a contract with Deusdona for the delivery of certain relics which the latter says he possesses. Eginhard makes no inquiry how he came by them; otherwise, the transaction is innocent enough.

Deusdona turns out to be a swindler, and has no relics. Thereupon Eginhard's agent, after due fasting and prayer, breaks open the tombs and helps himself.

Eginhard discovers by the self-betrayal of his brother abbot, Hildoin, that portions of his relics have been stolen and conveyed to the latter. With much ado he succeeds in getting them back.

Hildoin's agent, Hunus, in delivering these stolen goods to him, at first declared they were the relics of St. Tiburtius, which Hildoin desired him to obtain; but afterwards invented a story of their being the product of a theft, which the providential drowsiness of his companions enabled him to perpetrate, from the relics which Hildoin well knew were the property of his friend.

Lunison, on the contrary, swears that all this story is false, and that he himself was bribed by Hunus to allow him to steal what he pleased from the property confided to his own and his brother's care by their guest Ratleig. And the honest notary himself seems to have no hesitation about lying and stealing to any extent, where the acquisition of relics is the object in view.

For a parallel to these transactions one must
read a police report of the doings of a " long firm "
or of a set of horse-coupers ; yet Eginhard seems
to be aware of nothing, but that he has been
rather badly used by his friend Hildoin, and the
" nequissimus nebulo " Hunus.

It is not easy for a modern Protestant, still less
for any one who has the least tincture of scientific
culture, whether physical or historical, to picture
to himself the state of mind of a man of the
ninth century, however cultivated, enlightened,
and sincere he may have been. His deepest convictions, his most cherished hopes, were bound up
with the belief in the miraculous. Life was a
constant battle between saints and demons for the
possession of the souls of men. The most superstitious among our modern countrymen turn to
supernatural agencies only when natural causes
seem insufficient ; to Eginhard and his friends the
supernatural was the rule ; and the sufficiency of
natural causes was allowed only when there was
nothing to suggest others.

Moreover, it must be recollected that the
possession of miracle-working relics was greatly
coveted, not only on high, but on very low
grounds. To a man like Eginhard, the mere
satisfaction of the religious sentiment was
obviously a powerful attraction. But, more than
this, the possession of such a treasure was an
immense practical advantage. If the saints were

duly flattered and worshipped, there was no
telling what benefits might result from their
interposition on your behalf. For physical evils,
access to the shrine was like the grant of the use
of a universal pill and ointment manufactory;
and pilgrimages thereto might suffice to cleanse
the performers from any amount of sin. A letter
to Lupus, subsequently Abbot of Ferrara, written
while Eginhard was smarting under the grief
caused by the loss of his much-loved wife Imma,
affords a striking insight into the current view of
the relation between the glorified saints and their
worshippers. The writer shows that he is anything but satisfied with the way in which he has
been treated by the blessed martyrs whose remains he has taken such pains to "convey" to
Seligenstadt, and to honour there as they would
never have been honoured in their Roman obscurity.

> It is an aggravation of my grief and a reopening of my wound,
> that our vows have been of no avail, and that the faith which
> we placed in the merits and intervention of the martyrs has
> been utterly disappointed.

We may admit, then, without impeachment of
Eginhard's sincerity, or of his honour under all
ordinary circumstances, that when piety, self-
interest, the glory of the Church in general, and
that of the church at Seligenstadt in particular,
all pulled one way, even the workaday principles
of morality were disregarded; and, *a fortiori*,

anything like proper investigation of the reality
of alleged miracles was thrown to the winds.

And if this was the condition of mind of such a
man as Eginhard, what is it not legitimate to
suppose may have been that of Deacon Deusdona,
Lunison, Hunus, and Company, thieves and cheats
by their own confession, or of the probably
hysterical nun, or of the professional beggars, for
whose incapacity to walk and straighten themselves there is no guarantee but their own? Who
is to make sure that the exorcist of the demon
Wiggo was not just such another priest as Hunus;
and is it not at least possible, when Eginhard's
servants dreamed, night after night, in such a
curiously coincident fashion, that a careful inquirer
might have found they were very anxious to
please their master?

Quite apart from deliberate and conscious
fraud (which is a rarer thing than is often
supposed), people, whose mythopœic faculty is
once stirred, are capable of saying the thing that
is not, and of acting as they should not, to an
extent which is hardly imaginable by persons
who are not so easily affected by the contagion of
blind faith. There is no falsity so gross that
honest men and, still more, virtuous women,
anxious to promote a good cause, will not lend
themselves to it without any clear consciousness
of the moral bearings of what they are doing.

The cases of miraculously-effected cures of

the shrine of his favourite saints could be increased
by such a procedure.   There is no impeachment
of his honour in the supposition.   The logic of the
matter is quite simple, if somewhat sophistical.
The holiness of the church of the martyrs guarantees the reality of the appearance of the Archangel
Gabriel there ; and what the archangel says must
be true.   Therefore, if anything seem to be wrong,
that must be the mistake of the transmitter ; and,
in justice to the archangel, it must be suppressed
or set right.   This sort of "reconciliation" is not
unknown in quite modern times, and among people
who would be very much shocked to be compared
with a "benighted papist" of the ninth century.

The readers of this essay are, I imagine, very
largely composed of people who would be shocked
to be regarded as anything but enlightened
Protestants.   It is not unlikely that those of
them who have accompanied me thus far may be
disposed to say, " Well, this is all very amusing as
a story, but what is the practical interest of it?
We are not likely to believe in the miracles worked
by the spolia of SS. Marcellinus and Petrus, or by
those of any other saints in the Roman Calendar."

The practical interest is this : if you do not believe
in these miracles recounted by a witness whose
character and competency are firmly established,
whose sincerity cannot be doubted, and who
appeals to his sovereign and other contemporaries
as witnesses of the truth of what he says, in a

document of which a MS. copy exists, probably
dating within a century of the author's death,
why do you profess to believe in stories of a like
character, which are found in documents of the
dates and of the authorship of which nothing
is certainly determined, and no known copies of
which come within two or three centuries of the
events they record ?   If it be true that the four
Gospels and the Acts were written by Matthew,
Mark, Luke, and John, all that we know of these
persons comes to nothing in comparison with our
knowledge of Eginhard ; and not only is there no
proof that the traditional authors of these works
wrote them, but very strong reasons to the contrary
may be alleged.   If, therefore, you refuse to believe
that "Wiggo" was cast out of the possessed girl
on Eginhard's authority, with what justice can you
profess to believe that the legion of devils were
cast out of the man among the tombs of the
Gadarenes ?   And if, on the other hand, you accept
Eginhard's evidence, why do you laugh at the
supposed efficacy of relics and the saint-worship of
the modern Romanists ?   It cannot be pretended,
in the face of all evidence, that the Jews of the
year 30 A.D., or thereabouts, were less imbued
with the belief in the supernatural than were the
Franks of the year 800 A.D.   The same influences
were at work in each case, and it is only reasonable
to suppose that the results were the same.   If the
evidence of Eginhard is insufficient to lead reason-

trustworthy interpreter of their significance. When a man testifies to a miracle, he not only states a fact, but he adds an interpretation of the fact. We may admit his evidence as to the former, and yet think his opinion as to the latter worthless. If Eginhard's calm and objective narrative of the historical events of his time is no guarantee for the soundness of his judgment where the supernatural is concerned, the heated rhetoric of the Apostle of the Gentiles, his absolute confidence in the "inner light," and the extraordinary conceptions of the nature and requirements of logical proof which he betrays, in page after page of his Epistles, afford still less security.

There is a comparatively modern man wno shared to the full Paul's trust in the "inner light," and who, though widely different from the fiery evangelist of Tarsus in various obvious particulars, yet, if I am not mistaken, shares his deepest characteristics. I speak of George Fox, who separated himself from the current Protestantism of England, in the seventeenth century, as Paul separated himself from the Judaism of the first century, at the bidding of the "inner light"; who went through persecutions as serious as those which Paul enumerates; who was beaten, stoned, cast out for dead, imprisoned nine times, sometimes for long periods; who was in perils on land and perils at sea. George Fox was an even more widely-travelled missionary; while his success in founding congregations, and his energy in visiting them, not merely in Great Britain and Ireland and the West India Islands, but on the continent of Europe and that of North America, were no less remarkable. A few years after Fox began to preach, there were reckoned to be a thousand Friends in prison in the various gaols of England; at his death, less than fifty years after the foundation of the sect, there were 70,000 Quakers in the United Kingdom. The cheerfulness with which these people—women as well as men—underwent martyrdom in this country and in the New England States is one of the most remarkable facts in the history of religion.

No one who reads the voluminous autobiography of " Honest George " can doubt the man's utter truthfulness; and though, in his multitudinous letters, he but rarely rises far above the incoherent commonplaces of a street preacher, there can be no question of his power as a speaker, nor any doubt as to the dignity and attractiveness of his personality, or of his possession of a large amount of practical good sense and governing faculty.

But that George Fox had full faith in his own powers as a miracle-worker, the following passage of his autobiography (to which others might be added) demonstrates :—

Now after I was set at liberty from Nottingham gaol (where I had been kept a prisoner a pretty long time) I travelled as

before, in the work of the Lord. And coming to Mansfield
Woodhouse, there was a distracted woman, under a doctor's
hand, with her hair let loose all about her ears; and he was
about to let her blood, she being first bound, and many people
being about her, holding her by violence; but he could get no
blood from her. And I desired them to unbind her and let her
alone; for they could not touch the spirit in her by which she
was tormented. So they did unbind her, and I was moved to
speak to her, and in the name of the Lord to bid her be quiet
and still. And she was so. And the Lord's power settled her
mind and she mended; and afterwards received the truth and
continued in it to her death. And the Lord's name was
honoured; to whom the glory of all His works belongs. Many
great and wonderful things were wrought by the heavenly power
in those days. For the Lord made bare His omnipotent arm and
manifested His power to the astonishment of many; by the
healing virtue whereof many have been delivered from great
infirmities, and the devils were made subject through His name :
of which particular instances might be given beyond what this
unbelieving age is able to receive or bear.[1]

It needs no long study of Fox's writings, however, to arrive at the conviction that the distinction between subjective and objective verities had
not the same place in his mind as it has in that of
an ordinary mortal. When an ordinary person would
say "I thought so and so," or "I made up my
mind to do so and so," George Fox says, "It was
opened to me," or "at the command of God I
did so and so." "Then at the command of God, on
the ninth day of the seventh month 1643 (Fox being
just nineteen), I left my relations and brake off all

[1] *A Journal or Historical Account of the Life, Travels,
Sufferings, and Christian Experiences, &c., of George Fox.* Ed.
1694, pp. 27, 28.

familiarity or friendship with young or old."
"About the beginning of the year 1647 I was
moved of the Lord to go into Darbyshire." Fox
hears voices and he sees visions, some of which he
brings before the reader with apocalyptic power in
the simple and strong English, alike untutored
and undefiled, of which, like John Bunyan, his
contemporary, he was a master.

"And one morning, as I was sitting by the fire,
a great cloud came over me and a temptation beset
me; and I sate still. And it was said, *All things
come by Nature.* And the elements and stars came
over me; so that I was in a manner quite clouded
with it. . . . And as I sate still under it, and let it
alone, a living hope arose in me, and a true voice
arose in me which said, *There is a living God who
made all things.* And immediately the cloud and
the temptation vanished away, and life rose over
it all, and my heart was glad and I praised the
living God " (p. 13).

If George Fox could speak, as he proves in this
and some other passages he could write, his astounding influence on the contemporaries of Milton
and of Cromwell is no mystery. But this modern
reproduction of the ancient prophet, with his
"Thus saith the Lord," "This is the work of the
Lord," steeped in supernaturalism and glorying in
blind faith, is the mental antipodes of the philosopher, founded in naturalism and a fanatic for
evidence, to whom these affirmations inevitably

suggest the previous question: "How do you know that the Lord saith it?" "How do you know that the Lord doeth it?" and who is compelled to demand that rational ground for belief, without which, to the man of science, assent is merely an immoral pretence.

And it is this rational ground of belief which the writers of the Gospels, no less than Paul, and Eginhard, and Fox, so little dream of offering that they would regard the demand for it as a kind of blasphemy.

VI

POSSIBILITIES AND IMPOSSIBILITIES

[1891]

In the course of a discussion which has been going on during the last two years,[1] it has been maintained by the defenders of ecclesiastical Christianity that the demonology of the books of the New Testament is an essential and integral part of the revelation of the nature of the spiritual world promulgated by Jesus of Nazareth. Indeed, if the historical accuracy of the Gospels and of the Acts of the Apostles is to be taken for granted, if the teachings of the Epistles are divinely inspired, and if the universal belief and practice of the primitive Church are the models which all later times must follow, there can be no doubt that those who accept the demonology are in the right. It is as plain as language can make it, that the writers of the Gospels believed in the existence

---

[1] 1889–1891. See the next Essay (VII) and those which follow it.

of Satan and the surbordinate ministers of evil as
strongly as they believed in that of God and the
angels, and that they had an unhesitating faith in
possession and in exorcism.   No reader of the first
three Gospels can hesitate to admit that, in the
opinion of those persons among whom the traditions out of which they are compiled arose, Jesus
held, and constantly acted upon, the same theory
of the spiritual world.   Nowhere do we find the
slightest hint that he doubted the theory, or
questioned the efficacy of the curative operations
based upon it.

Thus, when such a story as that about the
Gadarene swine is placed before us, the importance
of the decision, whether it is to be accepted or
rejected, cannot be overestimated.   If the demonological part of it is to be accepted, the authority
of Jesus is unmistakably pledged to the demonological system current in Judæa in the first
century.   The belief in devils who possess men
and can be transferred from men to pigs, becomes
as much a part of Christian dogma as any article
of the creeds.   If it is to be rejected, there are two
alternative conclusions.   Supposing the Gospels to
be historically accurate, it follows that Jesus
shared in the errors, respecting the nature of the
spiritual world, prevalent in the age in which he
lived and among the people of his nation.   If, on
the other hand, the Gospel traditions gives us only
a popular version of the sayings and doings of

Jesus, falsely coloured and distorted by the superstitious imaginings of the minds through which it had passed, what guarantee have we that a similar unconscious falsification, in accordance with preconceived ideas, may not have taken place in respect of other reported sayings and doings? What is to prevent a conscientious inquirer from finding himself at last in a purely agnostic position with respect to the teachings of Jesus, and consequently with respect to the fundamentals of Christianity?

In dealing with the question whether the Gadarene story was to be believed or not, I confined myself altogether to a discussion of the value of the evidence in its favour. And, as it was easy to prove that this consists of nothing more than three partially discrepant, but often verbally coincident, versions of an original, of the authorship of which nobody knows anything, it appeared to me that it was wholly worthless. Even if the event described had been probable, such evidence would have required corroboration; being grossly improbable, and involving acts questionable in their moral and legal aspect, the three accounts sank to the level of mere tales.

Thus far, I am unable, even after the most careful revision, to find any flaw in my argument; and I incline to think none has been found by my critics—at least, if they have, they have kept the discovery to themselves.

In another part of my treatment of the case I have been less fortunate. I was careful to say that, for anything I could "absolutely prove to the contrary," there might be in the universe demonic beings who could enter into and possess men, and even be transferred from them to pigs; and that I, for my part, could not venture to declare *a priori* that the existence of such entities was "impossible." I was, however, no less careful to remark that I thought the evidence hitherto adduced in favour of the existence of such beings "ridiculously insufficient" to warrant the belief in them.

To my surprise, this statement of what, after the closest reflection, I still conceive to be the right conclusion, has been hailed as a satisfactory admission by opponents, and lamented as a perilous concession by sympathisers. Indeed, the tone of the comments of some candid friends has been such that I began to suspect that I must be entering upon a process of retrogressive metamorphosis which might eventually give me a place among the respectabilities. The prospect, perhaps, ought to have pleased me ; but I confess I felt something of the uneasiness of the tailor who said that, whenever a customer's circumference was either much less, or much more, than at the last measurement, he at once sent in his bill ; and I was not consoled until I recollected that, thirteen years ago, in discussing Hume's essay on

o 2

"Miracles," I had quoted, with entire assent, the following passage from his writings: "Whatever is intelligible and can be distinctly conceived implies no contradiction, and can never be proved false by any demonstrative argument or abstract reasoning *a priori.*"[1]

Now, it is certain that the existence of demons can be distinctly conceived. In fact, from the earliest times of which we have any record to the present day, the great majority of mankind have had extremely distinct conceptions of them, and their practical life has been more or less shaped by those conceptions. Further, the notion of the existence of such beings " implies no contradiction." No doubt, in our experience, intelligence and volition are always found in connection with a certain material organisation, and never disconnected with it; while, by the hypothesis, demons have no such material substratum. But then, as everybody knows, the exact relation between mental and physical phenomena, even in ourselves, is the subject of endless dispute. We may all have our opinions as to whether mental phenomena have a substratum distinct from that which is assumed to underlie material phenomena, or not; though if any one thinks he has demonstrative evidence of either the existence or the non-existence of a "soul," all I can say is, his notion of

---

[1] *Inquiry Concerning the Human Understanding*, p. 5; 1748. The passage is cited and discussed in my *Hume*, pp. 132, 133.

demonstration differs from mine. But, if it be impossible to demonstrate the non-existence of a "substance" of mental phenomena—that is, of a soul—independent of material "substance"; if the idea of such a "soul" is "intelligible and can be distinctly conceived," then it follows that it is not justifiable to talk of demons as "impossibilities." The idea of their existence implies no more "contradiction" than does the idea of the existence of pathogenic microbes in the air. Indeed, the microbes constitute a tolerably exact physical analogue of the "powers of the air" of ancient belief.

Strictly speaking, I am unaware of any thing that has a right to the title of an "impossibility" except a contradiction in terms. There are impossibilities logical, but none natural. A "round square," a "present past," "two parallel lines that intersect," are impossibilities, because the ideas denoted by the predicates, *round, present, intersect,* are contradictory of the ideas denoted by the subjects, *square, past, parallel.* But walking on water, or turning water into wine, or procreation without male intervention, or raising the dead, are plainly not "impossibilities" in this sense.

In the affirmation, that a man walked upon water, the idea of the subject is not contradictory of that in the predicate. Naturalists are familiar with insects which walk on water, and imagination has no more difficulty in putting a man in place of the insect than it has in giving a man some of the attributes of a bird and making an angel of him; or in ascribing to him the ascensive tendencies of a balloon, as the " levitationists" do.   Undoubtedly, there are very strong physical and biological arguments for thinking it extremely improbable that a man could be supported on the surface of the water as the insect is; or that his organisation could be compatible with the possession and use of wings; or that he could rise through the air without mechanical aid.   Indeed, if we have any reason to believe that our present knowledge of the nature of things exhausts the possibilities of nature, we might properly say that the attributes of men are contradictory of walking on water, or floating in the air, and consequently that these acts are truly "impossible" for him.   But it is sufficiently obvious, not only that we are at the beginning of our knowledge of nature, instead of having arrived at the end of it, but that the limitations of our faculties are such that we never can be in a position to set bounds to the possibilities of nature.   We have knowledge of what is happening and of what has happened; of what will happen we have and can have no more than expectation, grounded on our more or less correct reading of past experience and prompted by the faith, begotten of that experience, that the order of nature in the future will resemble its order in the past.

The same considerations apply to the other

examples of supposed miraculous events. The change of water into wine undoubtedly implies a contradiction, and is assuredly "impossible," if we are permitted to assume that the "elementary bodies" of the chemists are, now and for ever, immutable. Not only, however, is a negative proposition of this kind incapable of proof, but modern chemistry is inclining towards the contrary doctrine. And if carbon can be got out of hydrogen or oxygen, the conversion of water into wine comes within range of scientific possibility—it becomes a mere question of molecular arrangement.

As for virgin procreation, it is not only clearly imaginable, but modern biology recognises it as an every-day occurrence among some groups of animals. So with restoration to life after death. Certain animals, long as dry as mummies, and, to all appearance, as dead, when placed in proper conditions resume their vitality. It may be said that these creatures are not dead, but merely in a condition of suspended vitality. That, however, is only begging the question by making the incapacity for restoration to life part of the definition of death. In the absence of obvious lesions of some of the more important organs, it is no easy matter, even for experts, to say that an apparently dead man is incapable of restoration to life; and, in the recorded instances of such restoration, the want of any conclusive evidence that the man was dead is even more remarkable than the insufficiency of the testimony as to his coming to life again.

It may be urged, however, that there is, at any rate, one miracle certified by all three of the Synoptic Gospels which really does " imply a contradiction," and is, therefore, " impossible " in the strictest sense of the word. This is the well-known story of the feeding of several thousand men, to the complete satisfaction of their hunger, by the distribution of a few loaves and fishes among them; the wondrousness of this already somewhat surprising performance being intensified by the assertion that the quantity of the fragments of the meal, left over, amounted to much more than the original store.

Undoubtedly, if the operation is stated in its most general form; if it is to be supposed that a certain quantity, or magnitude, was divided into many more parts than the whole contained ; and that, after the subtraction of several thousands of such parts, the magnitude of the remainder amounted to more than the original magnitude, there does seem to be an *a priori* difficulty about accepting the proposition, seeing that it appears to be contradictory of the senses which we attach to the words " whole " and " parts " respectively. But this difficulty is removed if we reflect that we are not, in this case, dealing with magnitude in the abstract, or with " whole " and " parts " in their mathematical sense, but with concrete things, many of which are known to possess the power of growing, or increasing in magnitude. They thus furnish us with a conception of growth which we may, in imagination, apply to loaves and fishes ; just as we may, in imagination, apply the idea of wings to the idea of a man. It must be admitted that a number of sheep might be fed on a pasture, and yet there might be more grass on the pasture, when the sheep left it, than there was at first. We may generalise this and other such facts into a perfectly definite conception of the increase of food in excess of consumption ; which thus becomes a possibility, the limitations of which are to be discovered only by experience. Therefore, if it is asserted that cooked food has been made to grow in excess of rapid consumption, that statement cannot logically be rejected as an *à priori* impossibility, however improbable experience of the capabilities of cooked food may justify us in holding it to be.

On the strength of this undeniable improbability, however, we not only have a right to demand, but are morally bound to require, strong evidence in its favour before we even take it into serious consideration. But what is the evidence in this case ? It is merely that of those three books,[1] which also concur in testifying to the truth

[1] The story in John vi. 5-14 is obviously derived from the "five thousand" narrative of the Synoptics.

of the monstrous legend of the herd of swine. In these three books, there are five accounts of a " miraculous feeding," which fall into two groups. Three of the stories, obviously derived from some common source, state that five loaves and two fishes sufficed to feed five thousand persons, and that twelve baskets of fragments remained over. In the two others, also obviously derived from a common source, distinct from the preceding, seven loaves and a few small fishes are distributed to four thousand persons, and seven baskets of fragments are left.

If we were dealing with secular records, I suppose no candid and competent student of history would entertain much doubt that the originals of the three stories and of the two are themselves merely divergent versions of some primitive story which existed before the three Synoptic gospels were compiled out of the body of traditions current about Jesus. This view of the case, however, is incompatible with a belief in the historical accuracy of the first and second gospels.[1] For these agree in making Jesus himself speak of both the " four thousand " and the " five thousand " miracle. " When I brake the five loaves among the five thousand, how many baskets full of broken pieces took ye up ? They say unto him, twelve. And when the seven among the four

---

[1] Matthew xvi. 5--12 ; Mark viii. 14-21.

thousand, how many baskets full of broken pieces took ye up? And they say unto him, seven."

Thus we are face to face with a dilemma the way of escape from which is not obvious. Either the "four thousand" and the "five thousand" stories are both historically true, and describe two separate events; or the first and second gospels testify to the very words of a conversation between Jesus and his disciples which cannot have been uttered.

My choice between these alternatives is determined by no *a priori* speculations about the possibility or impossibility of such events as the feeding of the four or of the five thousand. But I ask myself the question, What evidence ought to be produced before I could feel justified in saying that I believed such an event to have occurred? That question is very easily answered. Proof must be given (1) of the weight of the loaves and fishes at starting; (2) of the distribution to 4-5,000 persons, without any additional supply, of this quantity and quality of food; (3) of the satisfaction of these people's appetites; (4) of the weight and quality of the fragments gathered up into the baskets. Whatever my present notions of probability and improbability may be, satisfactory testimony under these four heads would lead me to believe that they were erroneous; and I should accept the so-called miracle as a new and unexpected example of the possibilities of nature

But when, instead of such evidence, nothing is produced but two sets of discrepant stories, originating nobody knows how or when, among persons who could believe as firmly in devils which enter pigs, I confess that my feeling is one of astonishment that any one should expect a reasonable man to take such testimony seriously.

I am anxious to bring about a clear understanding of the difference between "impossibilities" and "improbabilities," because mistakes on this point lay us open to the attacks of ecclesiastical apologists of the type of the late Cardinal Newman; acute sophists, who think it fitting to employ their intellects, as burglars employ dark lanterns for the discovery of other people's weak places, while they carefully keep the light away from their own position.

When it is rightly stated, the Agnostic view of "miracles" is, in my judgment, unassailable. We are *not* justified in the *a priori* assertion that the order of nature, as experience has revealed it to us, cannot change. In arguing about the miraculous, the assumption is illegitimate, because it involves the whole point in dispute. Furthermore, it is an assumption which takes us beyond the range of our faculties. Obviously, no amount of past experience can warrant us in anything more than a correspondingly strong expectation for the present and future. We find, practically, that

taken as true, in the latter as untrue; until something arises to modify the verdict, which, however
properly reached, may always be more or less
wrong, the best information being never complete,
and the best reasoning being liable to fallacy.

To quarrel with the uncertainty that besets us
in intellectual affairs, would be about as reasonable
as to object to live one's life, with due thought for
the morrow, because no man can be sure he will
be alive an hour hence. Such are the conditions
imposed upon us by nature, and we have to make
the best of them. And I think that the greatest
mistake those of us who are interested in the progress of free thought can make is to overlook these
limitations, and to deck ourselves with the dogmatic feathers which are the traditional adorn-
ment of our opponents. Let us be content with
rational certainty, leaving irrational certainties to
those who like to muddle their minds with them.
I cannot see my way to say that demons are impossibilities; but I am not more certain about anything, than I am that the evidence tendered in
favour of the demonology, of which the Gadarene
story is a typical example, is utterly valueless. I
cannot see my way to say that it is "impossible"
that the hunger of thousands of men should be
satisfied out of the food supplied by half-a-dozen
loaves and a fish or two; but it seems to me monstrous that I should be asked to believe it on the
faith of the five stories which testify to such an

occurrence. It is true that the position that
miracles are "impossible" cannot be sustained.
But I know of nothing which calls upon me
to qualify the grave verdict of Hume : "There
is not to be found, in all history, any
miracle attested by a sufficient number of men,
of such unquestioned goodness, education, and
learning as to secure us against all delusion in
themselves; of such undoubted integrity as to
place them beyond all suspicion of any design to
deceive others; of such credit and reputation in
the eyes of mankind as to have a great deal
to lose in case of their being detected in any
falsehood; and at the same time attesting facts,
performed in such a public manner, and in so
celebrated a part of the world, as to render the
detection unavoidable : *all which circumstances
are requisite to give us a full assurance in the testimony of men.*" [1]

The preceding paper called forth the following criticism signed
"Agnosco," to which I append my reply :—

WHILE agreeing generally with Professor Huxley's remarks
respecting miracles, in "The Agnostic Annual for 1892," it has
seemed to me that one of his arguments at least requires quali-
fication. The Professor, in maintaining that so-called miraculous
events are possible, although the evidence adduced is not
sufficient to render them probable, refers to the possibility of
changing water into wine by molecular re-composition. He
tells us that, "if carbon can be got out of hydrogen or oxygen,
the conversion of water into wine comes within range of scientific
possibility." But in maintaining that miracles (so-called) have

---

[1] Hume, *Inquiry*, sec. x., part ii.

a *prospective* possibility, Professor Huxley loses sight—at least, so it appears to me—of the question of their *retrospective* possibility. For, if it requires a certain degree of knowledge and experience, yet far from having been attained, to perform those acts which have been called miraculous, it is not only improbable, but impossible likewise, that they should have been done by men whose knowledge and experience were considerably less than our own. It has seemed to me, in fact, that this question of the retrospective possibility of miracles is more important to us Rationalists, and, for the matter of that, to Christians also, than the question of their prospective possibility, with which Professor Huxley's article mainly deals. Perhaps the Professor himself could help those of us who think so, by giving us his opinion.

I AM not sure that I fully appreciate the point raised by "Agnosco," nor the distinction between the prospective and the retrospective "possibility" of such a miracle as the conversion of water into wine. If we may contemplate such an event as "possible" in London in the year 1900, it must, in the same sense, have been "possible" in the year 30 (or thereabouts) at Cana in Galilee. If I should live so long, I shall take great interest in the announcement of the performance of this operation, say, nine years hence ; and, if there is no objection raised by chemical experts, I shall accept the fact that the feat has been performed, without hesitation. But I shall have no more ground for believing the Cana story than I had before ; simply because the evidence in its favour will remain, for me, exactly where it is. Possible or impossible, that evidence is worth nothing. To leave the safe ground of "no evidence" for speculations about impossibilities, consequent upon the want of scientific knowledge of the supposed workers of miracles, appears to me to be a mistake ; especially in view of the orthodox contention that they possessed supernatural power and supernatural knowledge.   T. H. HUXLEY.

VII

AGNOSTICISM

[1889]

WITHIN the last few months, the public has received much and varied information on the subject of agnostics, their tenets, and even their future. Agnosticism exercised the orators of the Church Congress at Manchester.[1] It has been furnished with a set of "articles" fewer, but not less rigid, and certainly not less consistent than the thirty-nine ; its nature has been analysed, and its future severely predicted by the most eloquent of that prophetical school whose Samuel is Auguste Comte. It may still be a question, however, whether the public is as much the wiser as might be expected, considering all the trouble that has been taken to enlighten it. Not only are the three accounts of the agnostic position sadly out of harmony with one another, but I

[1] See the *Official Report of the Church Congress held at Manchester*, October 1888, pp. 253, 254.

propose to show cause for my belief that all three
must be seriously questioned by any one who
employs the term " agnostic " in the sense in
which it was originally used.　The learned
Principal of King's College, who brought the
topic of Agnosticism before the Church Congress,
took a short and easy way of settling the
business :—

> But if this be so, for a man to urge, as an escape from this
> article of belief, that he has no means of a scientific knowledge
> of the unseen world, or of the future, is irrelevant.　His differ-
> ence from Christians lies not in the fact that he has no know-
> ledge of these things, but that he does not believe the authority
> on which they are stated.　He may prefer to call himself an
> Agnostic ; but his real name is an older one—he is an infidel ;
> that is to say, an unbeliever.　The word infidel, perhaps, carries
> an unpleasant significance.　Perhaps it is right that it should.
> It is, and it ought to be, an unpleasant thing for a man to have to
> say plainly that he does not believe in Jesus Christ.[1]

So much of Dr. Wace's address either explicitly
or implicitly concerns me, that I take upon
myself to deal with it; but, in so doing, it must
be understood that I speak for myself alone.　I
am not aware that there is any sect of Agnostics ;

---

[1] [In this place and in the eleventh essay, there are references
to the late Archbishop of York which are of no importance to
my main argument, and which I have expunged because I desire
to obliterate the traces of a temporary misunderstanding with a
man of rare ability, candour, and wit, for whom I entertained a
great liking and no less respect.　I rejoice to think now of
the (then) Bishop's cordial hail the first time we met after our
little skirmish, " Well, is it to be peace or war ? "　I replied,
" A little of both."　But there was only peace when we parted,
and ever after.]

and if there be, I am not its acknowledged prophet or pope. I desire to leave to the Comtists the entire monopoly of the manufacture of imitation ecclesiasticism.

Let us calmly and dispassionately consider Dr. Wace's appreciation of agnosticism. The agnostic, according to his view, is a person who says he has no means of attaining a scientific knowledge of the unseen world or of the future; by which somewhat loose phraseology Dr. Wace presumably means the theological unseen world and future. I cannot think this description happy, either in form or substance, but for the present it may pass. Dr. Wace continues, that is not "his difference from Christians." Are there then any Christians who say that they know nothing about the unseen world and the future? I was ignorant of the fact, but I am ready to accept it on the authority of a professional theologian, and I proceed to Dr. Wace's next proposition.

The real state of the case, then, is that the agnostic "does not believe the authority" on which "these things" are stated, which authority is Jesus Christ. He is simply an old-fashioned "infidel" who is afraid to own to his right name. As "Presbyter is priest writ large," so is "agnostic" the mere Greek equivalent for the Latin "infidel." There is an attractive simplicity about this solution of the problem; and it has that advantage of being somewhat offensive to the

persons attacked, which is so dear to the less
refined sort of controversialist. The agnostic
says, " I cannot find good evidence that so and so
is true." " Ah," says his adversary, seizing his
opportunity, " then you declare that Jesus Christ
was untruthful, for he said so and so ; " a very
telling method of rousing prejudice. But suppose
that the value of the evidence as to what Jesus
may have said and done, and as to the exact
nature and scope of his authority, is just that
which the agnostic finds it most difficult to determine. If I venture to doubt that the Duke of
Wellington gave the command " Up, Guards, and
at 'em ! " at Waterloo, I do not think that even
Dr. Wace would accuse me of disbelieving the
Duke. Yet it would be just as reasonable to do
this as to accuse any one of denying what Jesus
said, before the preliminary question as to what
he did say is settled.

Now, the question as to what Jesus really said
and did is strictly a scientific problem, which is
capable of solution by no other methods than
those practised by the historian and the literary
critic. It is a problem of immense difficulty,
which has occupied some of the best heads in
Europe for the last century ; and it is only of late
years that their investigations have begun to con-
verge towards one conclusion.[1]

---

[1] Dr. Wace tells us, "It may be asked how far we can rely on
the accounts we possess of our Lord's teaching on these subjects."

That kind of faith which Dr. Wace describes and lauds is of no use here. Indeed, he himself takes pains to destroy its evidential value.

"What made the Mahommedan world? Trust and faith in the declarations and assurances of Mahommed. And what made the Christian world? Trust and faith in the declarations and assurances of Jesus Christ and His Apostles" (*l. c.* p. 253). The triumphant tone of this imaginary catechism leads me to suspect that its author has hardly appreciated its full import. Presumably, Dr. Wace regards Mahommed as an unbeliever, or, to use the term which he prefers, infidel; and considers that his assurances have given rise to a vast delusion which has led, and is leading, millions of men straight to everlasting punishment. And this being so, the "Trust and faith" which have "made the Mahommedan world," in just the same sense as they have

And he seems to think the question appropriately answered by the assertion that it "ought to be regarded as settled by M. Renan's practical surrender of the adverse case." I thought I knew M. Renan's works pretty well, but I have contrived to miss this "practical" (I wish Dr. Wace had defined the scope of that useful adjective) surrender. However, as Dr. Wace can find no difficulty in pointing out the passage of M. Renan's writings, by which he feels justified in making his statement, I shall wait for further enlightenment, contenting myself, for the present, with remarking that if M. Renan were to retract and do penance in Notre-Dame to-morrow for any contributions to Biblical criticism that may be specially his property, the main results of that criticism, as they are set forth in the works of Strauss, Baur, Reuss, and Volkmar, for example, would not be sensibly affected.

"made the Christian world," must be trust and
faith in falsehood. No man who has studied
history, or even attended to the occurrences of
everyday life, can doubt the enormous practical
value of trust and faith; but as little will he be
inclined to deny that this practical value has not
the least relation to the reality of the objects of
that trust and faith. In examples of patient
constancy of faith and of unswerving trust, the
"Acta Martyrum" do not excel the annals of
Babism.[1]

The discussion upon which we have now
entered goes so thoroughly to the root of the
whole matter; the question of the day is so
completely, as the author of "Robert Elsmere"
says, the value of testimony, that I shall offer no
apology for following it out somewhat in detail;
and, by way of giving substance to the argument,
I shall base what I have to say upon a case,
the consideration of which lies strictly within the
province of natural science, and of that particular
part of it known as the physiology and pathology
of the nervous system.

I find, in the second Gospel (chap. v.), a statement, to all appearance intended to have the
same evidential value as any other contained in

---

[1] [See De Gobineau, *Les Religions et les Philosophies dans
l'Asie Centrale*; and the recently published work of Mr. E. G.
Browne, *The Episode of the Bab.*]

that history. It is the well-known story of the devils who were cast out of a man, and ordered, or permitted, to enter into a herd of swine, to the great loss and damage of the innocent Gerasene, or Gadarene, pig owners. There can be no doubt that the narrator intends to convey to his readers his own conviction that this casting out and entering in were effected by the agency of Jesus of Nazareth; that, by speech and action, Jesus enforced this conviction; nor does any inkling of the legal and moral difficulties of the case manifest itself.

On the other hand, everything that I know of physiological and pathological science leads me to entertain a very strong conviction that the phenomena ascribed to possession are as purely natural as those which constitute small-pox; everything that I know of anthropology leads me to think that the belief in demons and demoniacal possession is a mere survival of a once universal superstition, and that its persistence, at the present time, is pretty much in the inverse ratio of the general instruction, intelligence, and sound judgment of the population among whom it prevails. Everything that I know of law and justice convinces me that the wanton destruction of other people's property is a misdemeanour of evil example. Again, the study of history, and especially of that of the fifteenth, sixteenth, and seventeenth centuries, leaves no shadow of doubt

on my mind that the belief in the reality of
possession and of witchcraft, justly based, alike
by Catholics and Protestants, upon this and innumerable other passages in both the Old and New
Testaments, gave rise, through the special influence of Christian ecclesiastics, to the most
horrible persecutions and judicial murders of
thousands upon thousands of innocent men,
women, and children. And when I reflect that
the record of a plain and simple declaration upon
such an occasion as this, that the belief in witchcraft and possession is wicked nonsense, would
have rendered the long agony of mediæval
humanity impossible, I am prompted to reject, as
dishonouring, the supposition that such declaration was withheld out of condescension to
popular error.

"Come forth, thou unclean spirit, out of the
man" (Mark v. 8),[1] are the words attributed to
Jesus. If I declare, as I have no hesitation in
doing, that I utterly disbelieve in the existence of
"unclean spirits," and, consequently, in the possibility of their "coming forth" out of a man, I
suppose that Dr. Wace will tell me I am
disregarding the testimony "of our Lord."
For, if these words were really used, the most
resourceful of reconcilers can hardly venture
to affirm that they are compatible with a disbelief "in these things." As the learned and

---

[1] Here, as always, the revised version is cited.

fair-minded, as well as orthodox, Dr. Alexander remarks, in an editorial note to the article "Demoniacs," in the "Biblical Cyclopædia" (vol. i. p. 664, note) :—

. . . On the lowest grounds on which our Lord and His Apostles can be placed they must, at least, be regarded as *honest* men. Now, though honest speech does not require that words should be used always and only in their etymological sense, it does require that they should not be used so as to affirm what the speaker knows to be false. Whilst, therefore, our Lord and His Apostles might use the word δαιμονίζεσθαι, or the phrase, δαιμόνιον εχειν, as a popular description of certain diseases, without giving in to the belief which lay at the source of such a mode of expression, they could not speak of demons entering into a man, or being cast out of him, without pledging them-selves to the belief of an actual possession of the man by the demons. (Campbell, *Prel. Diss.* vi. 1, 10.) If, consequently, they did not hold this belief, they spoke not as honest men.

The story which we are considering does not rest on the authority of the second Gospel alone. The third confirms the second, especially in the matter of commanding the unclean spirit to come out of the man (Luke viii. 29); and, although the first Gospel either gives a different version of the same story, or tells another of like kind, the essential point remains: "If thou cast us out, send us away into the herd of swine. And He said unto them : Go ! " (Matt. viii. 31, 32).

If the concurrent testimony of the three synoptics, then, is really sufficient to do away with all rational doubt as to a matter of fact of the utmost practical and speculative importance—

belief or disbelief in which may affect, and has affected, men's lives and their conduct towards other men, in the most serious way—then I am bound to believe that Jesus implicitly affirmed himself to possess a " knowledge of the unseen world," which afforded full confirmation of the belief in demons and possession current among his contemporaries. If the story is true, the mediæval theory of the invisible world may be, and probably is, quite correct; and the witch-finders, from Sprenger to Hopkins and Mather, are much-maligned men.

On the other hand, humanity, noting the frightful consequences of this belief; common sense, observing the futility of the evidence on which it is based, in all cases that have been properly investigated; science, more and more seeing its way to inclose all the phenomena of so-called "possession" within the domain of pathology, so far as they are not to be relegated to that of the police—all these powerful influences concur in warning us, at our peril, against accepting the belief without the most careful scrutiny of the authority on which it rests.

I can discern no escape from this dilemma: either Jesus said what he is reported to have said, or he did not. In the former case, it is inevitable that his authority on matters connected with the "unseen world" should be roughly shaken; in the latter, the blow falls upon the

authority of the synoptic Gospels. If their report
on a matter of such stupendous and far-reaching
practical import as this is untrustworthy, how can
we be sure of its trustworthiness in other cases?
The favourite " earth," in which the hard-pressed
reconciler takes refuge, that the Bible does not
profess to teach science,[1] is stopped in this
instance. For the question of the existence of
demons and of possession by them, though it lies
strictly within the province of science, is also of
the deepest moral and religious significance. If
physical and mental disorders are caused by demons, Gregory of Tours and his contemporaries
rightly considered that relics and exorcists were
more useful than doctors; the gravest questions
arise as to the legal and moral responsibilities of
persons inspired by demoniacal impulses; and our
whole conception of the universe and of our

---

[1] Does any one really mean to say that there is any internal or
external criterion by which the reader of a biblical statement, in
which scientific matter is contained, is enabled to judge whether
it is to be taken *au sérieux* or not? Is the account of the
Deluge. accepted as true in the New Testament, less precise and
specific than that of the call of Abraham, also accepted as true
therein? By what mark does the story of the feeding with
manna in the wilderness, which involves some very curious
scientific problems, show that it is meant merely for edification,
while the story of the inscription of the Law on stone by the
hand of Jahveh is literally true? If the story of the Fall is not
the true record of an historical occurrence, what becomes of
Pauline theology? Yet the story of the Fall as directly con-
flicts with probability, and is as devoid of trustworthy evidence,
as that of the Creation or that of the Deluge, with which it
forms an harmoniously legendary series.

relations to it becomes totally different from what it would be on the contrary hypothesis.

The theory of life of an average mediæval Christian was as different from that of an average nineteenth-century Englishman as that of a West African negro is now, in these respects. The modern world is slowly, but surely, shaking off these and other monstrous survivals of savage delusions; and, whatever happens, it will not return to that wallowing in the mire. Until the contrary is proved, I venture to doubt whether, at this present moment, any Protestant theologian, who has a reputation to lose, will say that he believes the Gadarene story.

The choice then lies between discrediting those who compiled the Gospel biographies and disbelieving the Master, whom they, simple souls, thought to honour by preserving such traditions of the exercise of his authority over Satan's invisible world. This is the dilemma. No deep scholarship, nothing but a knowledge of the revised version (on which it is to be supposed all that mere scholarship can do has been done), with the application thereto of the commonest canons of common sense, is needful to enable us to make a choice between its alternatives. It is hardly doubtful that the story, as told in the first Gospel, is merely a version of that told in the second and third. Nevertheless, the discrepancies are serious and irreconcilable; and, on this ground alone, a suspension of judgment, at the least, is called for. But there is a great deal more to be said. From the dawn of scientific biblical criticism until the present day, the evidence against the long-cherished notion that the three synoptic Gospels are the works of three independent authors, each prompted by Divine inspiration, has steadily accumulated, until, at the present time, there is no visible escape from the conclusion that each of the three is a compilation consisting of a groundwork common to all three— the threefold tradition; and of a superstructure, consisting, firstly, of matter common to it with one of the others, and, secondly, of matter special to each. The use of the terms "groundwork" and "superstructure" by no means implies that the latter must be of later date than the former. On the contrary, some parts of it may be, and probably are, older than some parts of the groundwork.[1]

The story of the Gadarene swine belongs to the groundwork; at least, the essential part of it, in which the belief in demoniac possession is expressed, does; and therefore the compilers of the first, second, and third Gospels, whoever they

[1] See, for an admirable discussion of the whole subject, Dr. Abbott's article on the Gospels in the *Encyclopædia Britannica ;* and the remarkable monograph by Professor Volkmar, *Jesus Nazarenus und die erste christliche Zeit* (1882). Whether we agree with the conclusions of these writers or not, the method of critical investigation which they adopt is unimpeachable.

were, certainly accepted that belief (which, indeed, was universal among both Jews and pagans at that time), and attributed it to Jesus.

What, then, do we know about the originator, or originators, of this groundwork—of that three-fold tradition which all three witnesses (in Paley's phrase) agree upon—that we should allow their mere statements to outweigh the counter arguments of humanity, of common sense, of exact science, and to imperil the respect which all would be glad to be able to render to their Master?

Absolutely nothing.[1]   There is no proof, nothing more than a fair presumption, that any one of the Gospels existed, in the state in which we find it in the authorised version of the Bible, before the second century, or, in other words, sixty or seventy years after the events recorded. And, between that time and the date of the oldest extant manuscripts of the Gospels, there is no telling what additions and alterations and interpolatións may have been made. It may be said that this is all mere speculation, but it is a good deal more.   As competent scholars and honest men, our revisers have felt compelled to point out that such things have happened even

---

[1] Notwithstanding the hard words shot at me from behind the hedge of anonymity by a writer in a recent number of the *Quarterly Review*, I repeat, without the slightest fear of refutation, that the four Gospels, as they have come to us, are the work of unknown writers.

since the date of the oldest known manuscripts.
The oldest two copies of the second Gospel end
with the 8th verse of the 16th chapter; the
remaining twelve verses are spurious, and it is
noteworthy that the maker of the addition has not
hesitated to introduce a speech in which Jesus
promises his disciples that "in My name shall
they cast out devils."

The other passage " rejected to the margin " is
still more instructive. It is that touching
apologue, with its profound ethical sense, of the
woman taken in adultery—which, if internal
evidence were an infallible guide, might well be
affirmed to be a typical example of the teachings
of Jesus. Yet, say the revisers, pitilessly, "Most
of the ancient authorities omit John vii. 53–viii.
11." Now let any reasonable man ask himself
this question. If, after an approximate settlement of the canon of the New Testament, and
even later than the fourth and fifth centuries,
literary fabricators had the skill and the audacity
to make such additions and interpolations as
these, what may they have done when no one
had thought of a canon; when oral tradition, still
unfixed, was regarded as more valuable than such
written records as may have existed in the latter
portion of the first century? Or, to take the
other alternative, if those who gradually settled
the canon did not know of the existence of the
oldest codices which have come down to us; or if,

knowing them, they rejected their authority, what is to be thought of their competency as critics of the text?

People who object to free criticism of the Christian Scriptures forget that they are what they are in virtue of very free criticism; unless the advocates of inspiration are prepared to affirm that the majority of influential ecclesiastics during several centuries were safeguarded against error. For, even granting that some books of the period were inspired, they were certainly few amongst many; and those who selected the canonical books, unless they themselves were also inspired, must be regarded in the light of mere critics, and, from the evidence they have left of their intellectual habits, very uncritical critics. When one thinks that such delicate questions as those involved fell into the hands of men like Papias (who believed in the famous millenarian grape story); of Irenæus with his "reasons" for the existence of only four Gospels; and of such calm and dispassionate judges as Tertullian, with his "Credo quia impossibile": the marvel is that the selection which constitutes our New Testament is as free as it is from obviously objectionable matter. The apocryphal Gospels certainly deserve to be apocryphal; but one may suspect that a little more critical discrimination would have enlarged the Apocrypha not inconsiderably.

At this point a very obvious objection arises

and deserves full and candid consideration. It may be said that critical scepticism carried to the length suggested is historical pyrrhonism; that if we are altogether to discredit an ancient or a modern historian, because he has assumed fabulous matter to be true, it will be as well to give up paying any attention to history. It may be said, and with great justice, that Eginhard's "Life of Charlemagne" is none the less trustworthy because of the astounding revelation of credulity, of lack of judgment, and even of respect for the eighth commandment, which he has unconsciously made in the "History of the Translation of the Blessed Martyrs Marcellinus and Paul." Or, to go no further back than the last number of the *Nineteenth Century*, surely that excellent lady, Miss Strickland, is not to be refused all credence, because of the myth about the second James's remains, which she seems to have unconsciously invented.

Of course this is perfectly true. I am afraid there is no man alive whose witness could be accepted, if the condition precedent were proof that he had never invented and promulgated a myth. In the minds of all of us there are little places here and there, like the indistinguishable spots on a rock which give foothold to moss or stonecrop; on which, if the germ of a myth fall, it is certain to grow, without in the least degree affecting our accuracy or truthfulness elsewhere. Sir Walter Scott knew that he could not repeat a

and deadly effects on both. For anything I can absolutely prove to the contrary, there may be spiritual things capable of the same transmigration, with like effects. Moreover I am bound to add that perfectly truthful persons, for whom I have the greatest respect, believe in stories about spirits of the present day, quite as improbable as that we are considering.

So I declare, as plainly as I can, that I am unable to show cause why these transferable devils should not exist; nor can I deny that, not merely the whole Roman Church, but many Wacean "infidels" of no mean repute, do honestly and firmly believe that the activity of such like demonic beings is in full swing in this year of grace 1889.

Nevertheless, as good Bishop Butler says, " probability is the guide of life ; " and it seems to me that this is just one of the cases in which the canon of credibility and testimony, which I have ventured to lay down, has full force. So that, with the most entire respect for many (by no means for all) of our witnesses for the truth of demonology, ancient and modern, I conceive their evidence on this particular matter to be ridiculously insufficient to warrant their conclusion.[1]

---

[1] Their arguments, in the long run, are always reducible to one form. Otherwise trustworthy witnesses affirm that such and such events took place. These events are inexplicable, except the agency of "spirits" is admitted. Therefore "spirits" were the cause of the phenomena.

And the heads of the reply are always the same. Remember

After what has been said, I do not think that
any sensible man, unless he happen to be angry,
will accuse me of " contradicting the Lord and His
Apostles " if I reiterate my total disbelief in the
whole Gadarene story. But, if that story is dis-
credited, all the other stories of demoniac possession fall under suspicion. And if the belief in
demons and demoniac possession, which forms the
sombre background of the whole picture of primitive Christianity, presented to us in the New
Testament, is shaken, what is to be said, in any
case, of the uncorroborated testimony of the
Gospels with respect to " the unseen world " ?

I am not aware that I have been influenced by
any more bias in regard to the Gadarene story
than I have been in dealing with other cases of
like kind the investigation of which has interested
me. I was brought up in the strictest school of
evangelical orthodoxy; and when I was old enough
to think for myself, I started upon my journey of
inquiry with little doubt about the general truth
of what I had been taught ; and with that feeling

Goethe's aphorism : " Alles factische ist schon Theorie." Trustworthy witnesses are constantly deceived, or deceive themselves,
in their interpretation of sensible phenomena. No one can
prove that the sensible phenomena, in these cases, could be
caused only bv the agency of spirits : and there is abundant
ground for believing that they may be produced in other ways.
Therefore, the utmost that can be reasonably asked for, on the
evidence as it stands, is suspension of judgment. And, on the
necessity for even that suspension, reasonable men may differ,
according to their views of probability.

of the unpleasantness of being called an "infidel" which, we are told, is so right and proper. Near my journey's end, I find myself in a condition of something more than mere doubt about these matters.

In the course of other inquiries, I have had to do with fossil remains which looked quite plain at a distance, and became more and more indistinct as I tried to define their outline by close inspection. There was something there—something which, if I could win assurance about it, might mark a new epoch in the history of the earth; but, study as long as I might, certainty eluded my grasp. So has it been with me in my efforts to define the grand figure of Jesus as it lies in the primary strata of Christian literature. Is he the kindly, peaceful Christ depicted in the Catacombs? Or is he the stern Judge who frowns above the altar of SS. Cosmas and Damianus? Or can he be rightly represented by the bleeding ascetic, broken down by physical pain, of too many mediæval pictures? Are we to accept the Jesus of the second, or the Jesus of the fourth Gospel, as the true Jesus? What did he really say and do; and how much that is attributed to him, in speech and action, is the embroidery of the various parties into which his followers tended to split themselves within twenty years of his death, when even the threefold tradition was only nascent?

If any one will answer these questions for me with something more to the point than feeble talk about the "cowardice of agnosticism," I shall be deeply his debtor. Unless and until they are satisfactorily answered, I say of agnosticism in this matter, "*J'y suis, et j'y reste.*"

But, as we have seen, it is asserted that I have no business to call myself an agnostic ; that, if I am not a Christian I am an infidel; and that I ought to call myself by that name of "unpleasant significance." Well, I do not care much what I am called by other people, and if I had at my side all those who, since the Christian era, have been called infidels by other folks, I could not desire better company. If these are my ancestors, I prefer, with the old Frank, to be with them wherever they are. But there are several points in Dr. Wace's contention which must be elucidated before I can even think of undertaking to carry out his wishes. I must, for instance, know what a Christian is. Now what is a Christian? By whose authority is the signification of that term defined ? Is there any doubt that the immediate followers of Jesus, the "sect of the Nazarenes," were strictly orthodox Jews differing from other Jews not more than the Sadducees, the Pharisees, and the Essenes differed from one another ; in fact, only in the belief that the Messiah, for whom the rest of their nation waited, had come ? Was not their chief, "James, the brother of the Lord," reverenced alike by Sadducee, Pharisee, and Nazarene? At the famous conference which, according to the Acts, took place at Jerusalem, does not James declare that "myriads" of Jews, who, by that time, had become Nazarenes, were "all zealous for the Law"? Was not the name of "Christian" first used to denote the converts to the doctrine promulgated by Paul and Barnabas at Antioch? Does the subsequent history of Christianity leave any doubt that, from this time forth, the "little rift within the lute" caused by the new teaching, developed, if not inaugurated, at Antioch, grew wider and wider, until the two types of doctrine irreconcilably diverged? Did not the primitive Nazarenism, or Ebionism, develop into the Nazarenism, and Ebionism, and Elkasaitism of later ages, and finally die out in obscurity and condemnation, as damnable heresy; while the younger doctrine throve and pushed out its shoots into that endless variety of sects, of which the three strongest survivors are the Roman and Greek Churches and modern Protestantism?

Singular state of things! If I were to profess the doctrine which was held by "James, the brother of the Lord," and by every one of the "myriads" of his followers and co-religionists in Jerusalem up to twenty or thirty years after the Crucifixion (and one knows not how much later at Pella), I should be condemned, with unanimity, as an ebionising heretic by the Roman, Greek, and

Protestant Churches! And, probably, this hearty
and unanimous condemnation of the creed, held by
those who were in the closest personal relation
with their Lord, is almost the only point upon
which they would be cordially of one mind. On
the other hand, though I hardly dare imagine
such a thing, I very much fear that the "pillars"
of the primitive Hierosolymitan Church would
have considered Dr. Wace an infidel. No one can
read the famous second chapter of Galatians and
the book of Revelation without seeing how narrow was even Paul's escape from a similar fate.
And, if ecclesiastical history is to be trusted, the
thirty-nine articles, be they right or wrong,
diverge from the primitive doctrine of the Nazarenes vastly more than even Pauline Christianity
did.

But, further than this, I have great difficulty
in assuring myself that even James, "the brother
of the Lord," and his "myriads" of Nazarenes,
properly represented the doctrines of their
Master. For it is constantly asserted by our
modern "pillars" that one of the chief features of
the work of Jesus was the instauration of Religion
by the abolition of what our sticklers for articles
and liturgies, with unconscious humour, call the
narrow restrictions of the Law. Yet, if James
knew this, how could the bitter controversy with
Paul have arisen; and why did not one or
the other side quote any of the various sayings of

Jesus, recorded in the Gospels, which directly bear on the question—sometimes, apparently, in opposite directions?

So, if I am asked to call myself an "infidel," I reply : To what doctrine do you ask me to be faithful ? Is it that contained in the Nicene and the Athanasian Creeds ? My firm belief is that the Nazarenes, say of the year 40, headed by James, would have stopped their ears and thought worthy of stoning the audacious man who propounded it to them. Is it contained in the so-called Apostles' Creed ? I am pretty sure that even that would have created a recalcitrant commotion at Pella in the year 70, among the Nazarenes of Jerusalem, who had fled from the soldiers of Titus And yet, if the unadulterated tradition of the teachings of "the Nazarene" were to be found anywhere, it surely should have been amidst those not very aged disciples who may have heard them as they were delivered.

Therefore, however sorry I may be to be unable to demonstrate that, if necessary, I should not be afraid to call myself an " infidel," I cannot do it. "Infidel" is a term of reproach, which Christians and Mahommedans, in their modesty, agree to apply to those who differ from them. If he had only thought of it, Dr. Wace might have used the term " miscreant," which, with the same etymological signification, has the advantage of being still more " unpleasant " to the persons to whom

it is applied. But why should a man be expected
to call himself a "miscreant" or an "infidel"?
That St. Patrick "had two birthdays because he
was a twin" is a reasonable and intelligible utterance beside that of the man who should declare
himself to be an infidel, on the ground of denying
his own belief. It may be logically, if not ethi-
cally, defensible that a Christian should call a
Mahommedan an infidel and *vice versâ;* but, on
Dr. Wace's principles, both ought to call themselves infidels, because each applies the term to
the other.

Now I am afraid that all the Mahommedan world
would agree in reciprocating that appellation to
Dr. Wace himself. I once visited the Hazar
Mosque, the great University of Mahommedanism,
in Cairo, in ignorance of the fact that I was un-
provided with proper authority. A swarm of
angry undergraduates, as I suppose I ought to
call them, came buzzing about me and my guide;
and if I had known Arabic, I suspect that "dog
of an infidel" would have been by no means the
most "unpleasant" of the epithets showered upon
me, before I could explain and apologise for the
mistake. If I had had the pleasure of Dr. Wace's
company on that occasion, the undiscriminative
followers of the Prophet would, I am afraid, have
made no difference between us; not even if they
had known that he was the head of an orthodox
Christian seminary. And I have not the smallest

doubt that even one of the learned mollahs, if his grave courtesy would have permitted him to say anything offensive to men of another mode of belief, would have told us that he wondered we did not find it " very unpleasant " to disbelieve in the Prophet of Islam.

From what precedes, I think it becomes sufficiently clear that Dr. Wace's account of the origin of the name of " Agnostic " is quite wrong. Indeed, I am bound to add that very slight effort to discover the truth would have convinced him that, as a matter of fact, the term arose otherwise. I am loath to go over an old story once more ; but more than one object which I have in view will be served by telling it a little more fully than it has yet been told.

Looking back nearly fifty years, I see myself as a boy, whose education has been interrupted, and who, intellectually, was left, for some years, altogether to his own devices. At that time, I was a voracious and omnivorous reader ; a dreamer and speculator of the first water, well endowed with that splendid courage in attacking any and every subject, which is the blessed compensation of youth and inexperience. Among the books and essays, on all sorts of topics from metaphysics to heraldry, which I read at this time, two left indelible impressions on my mind. One was Guizot's " History of Civilisation," the other was Sir William Hamilton's essay " On the Philosophy of the Unconditioned," which I came upon, by chance, in an odd volume of the "Edinburgh Review." The latter was certainly strange reading for a boy, and I could not possibly have understood a great deal of it;[1] nevertheless, I devoured it with avidity, and it stamped upon my mind the strong conviction that, on even the most solemn and important of questions, men are apt to take cunning phrases for answers; and that the limitation of our faculties, in a great number of cases, renders real answers to such questions, not merely actually impossible, but theoretically inconceivable.

Philosophy and history having laid hold of me in this eccentric fashion, have never loosened their grip. I have no pretension to be an expert in either subject; but the turn for philosophical and historical reading, which rendered Hamilton and Guizot attractive to me, has not only filled many lawful leisure hours, and still more sleepless ones, with the repose of changed mental occupation, but has not unfrequently disputed my proper work-time with my liege lady, Natural Science. In this way I have found it possible to cover a good deal of ground in the territory of philosophy; and all the more easily that I have never cared much about A's

---

[1] Yet I must somehow have laid hold of the pith of the matter, for, many years afterwards, when Dean Mansel's Bampton Lectures were published, it seemed to me I already knew all that this eminently agnostic thinker had to tell me.

or B's opinion's, but have rather sought to know
what answer he had to give to the questions I had
to put to him—that of the limitation of possible
knowledge being the chief. The ordinary examiner, with his "State the views of So-and-so,"
would have floored me at any time. If he had
said what do *you* think about any given problem,
I might have got on fairly well.

The reader who has had the patience to follow
the enforced, but unwilling, egotism of this
veritable history (especially if his studies have led
him in the same direction), will now see why my
mind steadily gravitated towards the conclusions
of Hume and Kant, so well stated by the
latter in a sentence, which I have quoted elsewhere.

" The greatest and perhaps the sole use of all
philosophy of pure reason is, after all, merely
negative, since it serves not as an organon for the
enlargement [of knowledge], but as a discipline for
its delimitation; and, instead of discovering
truth, has only the modest merit of preventing
error." [1]

When I reached intellectual maturity and
began to ask myself whether I was an atheist, a
theist, or a pantheist; a materialist or an idealist;
a Christian or a freethinker; I found that the
more I learned and reflected, the less ready was
the answer; until, at last, I came to the conclu-

[1] *Kritik der reinen Vernunft.* Edit. Hartenstein, p. 256.

sion that I had neither art nor part with any of these denominations, except the last. The one thing in which most of these good people were agreed was the one thing in which I differed from them. They were quite sure they had attained a certain "gnosis,"—had, more or less successfully, solved the problem of existence; while I was quite sure I had not, and had a pretty strong conviction that the problem was insoluble. And, with Hume and Kant on my side, I could not think myself presumptuous in holding fast by that opinion. Like Dante,

> Nel mezzo del cammin di nostra vita
> Mi ritrovai per una selva oscura,

but, unlike Dante, I cannot add,

> Che la diritta via era smarrita.

On the contrary, I had, and have, the firmest conviction that I never left the "verace via"—the straight road; and that this road led nowhere else but into the dark depths of a wild and tangled forest. And though I have found leopards and lions in the path; though I have made abundant acquaintance with the hungry wolf, that "with privy paw devours apace and nothing said," as another great poet says of the ravening beast; and though no friendly spectre has even yet offered his guidance, I was, and am, minded to go straight on, until I either come out on the other side of the

wood, or find there is no other side to it, at least,
none attainable by me.

This was my situation when I had the good
fortune to find a place among the members of that
remarkable confraternity of antagonists, long since
deceased, but of green and pious memory, the Metaphysical Society. Every variety of philosophical
and theological opinion was represented there, and
expressed itself with entire openness; most of my
colleages were -*ists* of one sort or another; and,
however kind and friendly they might be, I, the
man without a rag of a label to cover himself with,
could not fail to have some of the uneasy feelings
which must have beset the historical fox when,
after leaving the trap in which his tail remained,
he presented himself to his normally elongated
companions. So I took thought, and invented
what I conceived to be the appropriate title of
" agnostic." It came into my head as suggestively
antithetic to the " gnostic " of Church history, who
professed to know so much about the very things
of which I was ignorant; and I took the earliest
opportunity of parading it at our Society, to show
that I, too, had a tail, like the other foxes. To
my great satisfaction, the term took; and when
the *Spectator* had stood godfather to it, any
suspicion in the minds of respectable people, that
a knowledge of its parentage might have awakened
was, of course, completely lulled.

That is the history of the origin of the terms

" agnostic " and " agnosticism " ; and it will be observed that it does not quite agree with the confident assertion of the reverend Principal of King's
College, that " the adoption of the term agnostic is
only an attempt to shift the issue, and that it involves a mere evasion " in relation to the Church
and Christianity.[1]

The .last objection (I rejoice as much as my
readers must do, that it is the last) which I have
to take to Dr. Wace's deliverance before the Church
Congress arises, I am sorry to say, on a question of
morality.

" It is, and it ought to be," authoritatively declares this official representative of Christian
ethics, " an unpleasant thing for a man to have
to say plainly that he does not believe in Jesus
Christ " (l. c. p. 254).

Whether it is so depends, I imagine, a good deal
on whether the man was brought up in a Christian
household or not. I do not see why it should be
" unpleasant" for a Mahommedan or Buddhist to
say so. But that " it ought to be " unpleasant for
any man to say anything which he sincerely, and
after due deliberation, believes, is, to my mind, a
proposition of the most profoundly immoral
character. I verily believe that the great good
which has been effected in the world by Christianity has been largely counteracted by the pestilent

[1] *Report of the Church Congress,* Manchester, 1888, p. 252.

doctrine on which all the Churches have insisted, that honest disbelief in their more or less astonishing creeds is a moral offence, indeed a sin of the deepest dye, deserving and involving the same future retribution as murder and robbery. If we could only see, in one view, the torrents of hypocrisy and cruelty, the lies, the slaughter, the violations of every obligation of humanity, which have flowed from this source along the course of the history of Christian nations, our worst imaginations of Hell would pale beside the vision.

A thousand times, no! It ought *not* to be unpleasant to say that which one honestly believes or disbelieves. That it so constantly is painful to do so, is quite enough obstacle to the progress of mankind in that most valuable of all qualities, honesty of word or of deed, without erecting a sad concomitant of human weakness into something to be admired and cherished. The bravest of soldiers often, and very naturally, " feel it unpleasant " to go into action; but a court-martial which did its duty would make short work of the officer who promulgated the doctrine that his men *ought* to feel their duty unpleasant.

I am very well aware, as I suppose most thoughtful people are in these times, that the process of breaking away from old beliefs is extremely unpleasant; and I am much disposed to think that the encouragement, the consolation, and the peace afforded to earnest believers in even the

worst forms of Christianity are of great practical
advantage to them.   What deductions must be
made from this gain on the score of the harm done
to the citizen by the ascetic other-worldliness of
logical Christianity; to the ruler, by the hatred,
malice, and all uncharitableness of sectarian
bigotry; to the legislator, by the spirit of exclu-
siveness and domination of those that count themselves pillars of orthodoxy; to the philosopher, by
the restraints on the freedom of learning and
teaching which every Church exercises, when it is
strong enough; to the conscientious soul, by the
introspective hunting after sins of the mint and
cummin type, the fear of theological error, and the
overpowering terror of possible damnation, which
have accompanied the Churches like their shadow,
I need not now consider; but they are assuredly
not small.   If agnostics lose heavily on the one
side, they gain a good deal on the other.   People
who talk about the comforts of belief appear to
forget its discomforts; they ignore the fact that
the Christianity of the Churches is something
more than faith in the ideal personality of Jesus,
which they create for themselves, *plus* so much as
can be carried into practice, without disorganising
civil society, of the maxims of the Sermon on the
Mount.   Trip in morals or in doctrine (especially in
doctrine), without due repentance or retractation,
or fail to get properly baptized before you die, and
a *plébiscite* of the Christians of Europe, if they

were true to their creeds, would affirm your
everlasting damnation by an immense majority.

Preachers, orthodox and heterodox, din into our
ears that the world cannot get on without faith of
some sort.   There is a sense in which that is as
eminently as obviously true; there is another, in
which, in my judgment, it is as eminently as
obviously false, and it seems to me that the
hortatory, or pulpit, mind is apt to oscillate
between the false and the true meanings, without
being aware of the fact.

It is quite true that the ground of every one of
our actions, and the validity of all our reasonings,
rest upon the great act of faith, which leads us to
take the experience of the past as a safe guide in
our dealings with the present and the future.
From the nature of ratiocination, it is obvious that
the axioms, on which it is based, cannot be demonstrated by ratiocination.   It is also a trite observation that, in the business of life, we constantly
take the most serious action upon evidence of an
utterly insufficient character.   But it is surely
plain that faith is not necessarily entitled to
dispense with ratiocination because ratiocination
cannot dispense with faith as a starting-point;
and that because we are often obliged, by the
pressure of events, to act on very bad evidence, it
does not follow that it is proper to act on such
evidence when the pressure is absent.

The writer of the epistle to the Hebrews tells

us that "faith is the assurance of things hoped
for, the proving of things not seen." In the
authorised version, "substance" stands for
"assurance," and "evidence" for "proving."
The question of the exact meaning of the two
words, $\upsilon\pi o\sigma\tau a\sigma\iota\varsigma$ and $\H{\epsilon}\lambda\epsilon\gamma\chi o\varsigma$, affords a fine field
of discussion for the scholar and the metaphysician.
But I fancy we shall be not far from the mark if
we take the writer to have had in his mind the
profound psychological truth, that men constantly
feel certain about things for which they strongly
hope, but have no evidence, in the legal or logical
sense of the word; and he calls this feeling
"faith." I may have the most absolute faith that
a friend has not committed the crime of which he
is accused. In the early days of English history,
if my friend could have obtained a few more
compurgators of a like robust faith, he would have
been acquitted. At the present day, if I tendered
myself as a witness on that score, the judge would
tell me to stand down, and the youngest barrister
would smile at my simplicity. Miserable indeed
is the man who has not such faith in some of his
fellow-men—only less miserable than the man
who allows himself to forget that such faith is not,
strictly speaking, evidence; and when his faith is
disappointed, as will happen now and again, turns
Timon and blames the universe for his own
blunders. And so, if a man can find a friend, the
hypostasis of all his hopes, the mirror of his

ethical ideal, in the Jesus of any, or all, of the
Gospels, let him live by faith in that ideal. Who
shall or can forbid him ? But let him not delude
himself with the notion that his faith is evidence
of the objective reality of that in which he trusts.
Such evidence is to be obtained only by the use
of the methods of science, as applied to history
and to literature, and it amounts at present to
very little.

It appears that Mr. Gladstone some time ago
asked Mr. Laing if he could draw up a short
summary of the negative creed ; a body of
negative propositions, which have so far been
adopted on the negative side as to be what the
Apostles' and other accepted creeds are on the
positive ; and Mr. Laing at once kindly obliged
Mr. Gladstone with the desired articles—eight of
them.

If any one had preferred this request to me,
I should have replied that, if he referred to agnostics, they have no creed ; and, by the nature of
the case, cannot have any. Agnosticism, in fact,
is not a creed, but a method, the essence of which
lies in the rigorous application of a single principle.
That principle is of great antiquity ; it is as old as
Socrates ; as old as the writer who said, " Try all
things, hold fast by that which is good ; " it is the
foundation of the Reformation, which simply illustrated the axiom that every man should be able

to give a reason for the faith that is in him ; it is
the great principle of Descartes ; it is the fundamental axiom of modern science. Positively the
principle may be expressed : In matters of the
intellect, follow your reason as far as it will take
you, without regard to any other consideration.
And negatively : In matters of the intellect do
not pretend that conclusions are certain which are
not demonstrated or demonstrable. That I take
to be the agnostic faith, which if a man keep
whole and undefiled, he shall not be ashamed to
look the universe in the face, whatever the future
may have in store for him.

The results of the working out of the agnostic
principle will vary according to individual knowledge and capacity, and according to the general
condition of science. That which is unproven to-
day may be proven by the help of new discoveries
to-morrow. The only negative fixed points will
be those negations which flow from the demonstrable limitation of our faculties. And the only
obligation accepted is to have the mind always
open to conviction. Agnostics who never fail in
carrying out their principles are, I am afraid, as
rare as other people of whom the same consistency
can be truthfully predicated. But, if you were to
meet with such a phœnix and to tell him that you
had discovered that two and two make five, he
would patiently ask you to state your reasons for
that conviction, and express his readiness to

agree with you if he found them satisfactory. The apostolic injunction to " suffer fools gladly " should be the rule of life of a true agnostic. I am deeply conscious how far I myself fall short of this ideal, but it is my personal conception of what agnostics ought to be.

However, as I began by stating, I speak only for myself; and I do not dream of anathematizing and excommunicating Mr. Laing. But, when I consider his creed and compare it with the Athanasian, I think I have on the whole a clearer conception of the meaning of the latter. "Polarity," in Article VIII., for example, is a word about which I heard a good deal in my youth, when "Naturphilosophie" was in fashion, and greatly did I suffer from it. For many years past, whenever I have met with " polarity" anywhere but in a discussion of some purely physical topic, such as magnetism, I have shut the book. Mr. Laing must excuse me if the force of habit was too much for me when I read his eighth article.

And now, what is to be said to Mr. Harrison's remarkable deliverance " On the future of agnosticism " ? [1] I would that it were not my business to say anything, for I am afraid I can say nothing which shall manifest my great personal respect for this able writer, and for the zeal and energy with which he ever and anon galvanises the

[1] *Fortnightly Review,* Jan. 1889.

weakly frame of Positivism until it looks, more
than ever, like John Bunyan's Pope and Pagan
rolled into one. There is a story often repeated,
and I am afraid none the less mythical on that
account, of a valiant and loud-voiced corporal in
command of two full privates who, falling in with
a regiment of the enemy in the dark, orders it to
surrender under pain of instant annihilation by
his force; and the enemy surrenders accordingly.
I am always reminded of this tale when I read
the positivist commands to the forces of Christianity and of Science; only the enemy show no
more signs of intending to obey now than they
have done any time these forty years.

The allocution under consideration has a
certain papal flavour. Mr. Harrison speaks
with authority and not as one of the common scribes of the period. He knows not only
what agnosticism is and how it has come about,
but what will become of it. The agnostic is
to content himself with being the precursor of
the positivist. In his place, as a sort of navvy
levelling the ground and cleansing it of such
poor stuff as Christianity, he is a useful creature who deserves patting on the back, on condition that he does not venture beyond his
last. But let not these scientific Sanballats
presume that they are good enough to take part
in the building of the Temple—they are mere
Samaritans, doomed to die out in proportion as

the Religion of Humanity is accepted by mankind. Well, if that is their fate, they have time
to be cheerful. But let us hear Mr. Harrison's
pronouncement of their doom.

"Agnosticism is a stage in the evolution of
religion, an entirely negative stage, the point
reached by physicists, a purely mental conclusion,
with no relation to things social at all" (p. 154).
I am quite dazed by this declaration. Are there,
then, any "conclusions" that are not "purely
mental"? Is there "no relation to things social"
in "mental conclusions" which affect men's
whole conception of life? Was that prince of
agnostics, David Hume, particularly imbued with
physical science? Supposing physical science
to be non-existent, would not the agnostic
principle, applied by the philologist and the
historian, lead to exactly the same results? Is
the modern more or less complete suspension of
judgment as to the facts of the history of regal
Rome, or the real origin of the Homeric poems,
anything but agnosticism in history and in
literature? And if so, how can agnosticism be
the "mere negation of the physicist"?

"Agnosticism is a stage in the evolution of
religion." No two people agree as to what is
meant by the term "religion"; but if it means,
as I think it ought to mean, simply the reverence
and love for the ethical ideal, and the desire to
realise that ideal in life, which every man ought

to feel—then I say agnosticism has no more to do
with it than it has to do with music or painting.
If, on the other hand, Mr. Harrison, like most
people, means by "religion" theology, then, in my
judgment, agnosticism can be said to be a stage in
its evolution, only as death may be said to be
the final stage in the evolution of life.

> When agnostic logic is simply one of the canons of thought,
> agnosticism, as a distinctive faith, will have spontaneously
> disappeared (p. 155).

I can but marvel that such sentences as this,
and those already quoted, should have proceeded
from Mr. Harrison's pen. Does he really mean to
suggest that agnostics have a logic peculiar to
themselves? Will he kindly help me out of my
bewilderment when I try to think of "logic"
being anything else than the canon (which, I
believe, means rule) of thought? As to agnosticism being a distinctive faith, I have already
shown that it cannot possibly be anything of the
kind, unless perfect faith in logic is distinctive of
agnostics; which, after all, it may be.

> Agnosticism as a religious philosophy *per se* rests on an almost
> total ignoring of history and social evolution (p. 152).

But neither *per se* nor *per aliud* has agnosticism
(if I know anything about it) the least pretension
to be a religious philosophy; so far from resting
on ignorance of history, and that social evolution

of which history is the account, it is and has
been the inevitable result of the strict adherence
to scientific methods by historical investigators.
Our forefathers were quite confident about the
existence of Romulus and Remus, of King Arthur,
and of Hengist and Horsa. Most of us have
become agnostics in regard to the reality of these
worthies. It is a matter of notoriety of which
Mr. Harrison, who accuses us all so freely of
ignoring history, should not be ignorant, that the
critical process which has shattered the foundations of orthodox Christian doctrine owes its
origin, not to the devotees of physical science, but,
before all, to Richard Simon, the learned French
Oratorian, just two hundred years ago. I cannot
find evidence that either Simon, or any one of the
great scholars and critics of the eighteenth and
nineteenth centuries who have continued Simon's
work, had any particular acquaintance with
physical science. I have already pointed out
that Hume was independent of it. And certainly
one of the most potent influences in the same
direction, upon history in the present century, that
of Grote, did not come from the physical side.
Physical science, in fact, has had nothing directly
to do with the criticism of the Gospels; it is
wholly incompetent to furnish demonstrative
evidence that any statement made in these histories is untrue. Indeed, modern physiology can
find parallels in nature for events of apparently

thousand unwilling porters were once launched down the steep slopes of the fatal shore of Gennesaret.

The question of the place of religion as an element of human nature, as a force of human society, its origin, analysis, and functions, has never been considered at all from an agnostic point of view (p. 152).

I doubt not that Mr. Harrison knows vastly more about history than I do; in fact, he tells the public that some of my friends and I have had no opportunity of occupying ourselves with that subject. I do not like to contradict any statement which Mr. Harrison makes on his own authority; only, if I may be true to my agnostic principles, I humbly ask how he has obtained assurance on this head. I do not profess to know anything about the range of Mr. Harrison's studies; but as he has thought it fitting to start the subject, I may venture to point out that, on evidence adduced, it might be equally permissible to draw the conclusion that Mr. Harrison's other labours have not allowed him to acquire that acquaintance with the methods and results of physical science, or with the history of philosophy, or of philological and historical criticism, which is essential to any one who desires to obtain a right understanding of agnosticism. Incompetence in philosophy, and in all branches of science except mathematics, is the well-known mental characteristic of the founder of positivism. Faithfulness in disciples is an admirable quality in itself; the pity is that it not unfrequently leads to the imitation of the weaknesses as well as of the strength of the master. It is only such over-faithfulness which can account for a "strong mind really saturated with the historical sense" (p. 153) exhibiting the extraordinary forgetfulness of the historical fact of the existence of David Hume implied by the assertion that

it would be difficult to name a single known agnostic who has given to history anything like the amount of thought and study which he brings to a knowledge of the physical world (p. 153).

Whoso calls to mind what I may venture to term the bright side of Christianity—that ideal of manhood, with its strength and its patience, its justice and its pity for human frailty, its helpfulness to the extremity of self-sacrifice, its ethical purity and nobility, which apostles have pictured, in which armies of martyrs have placed their unshakable faith, and whence obscure men and women, like Catherine of Sienna and John Knox, have derived the courage to rebuke popes and kings—is not likely to underrate the importance of the Christian faith as a factor in human history, or to doubt that if that faith should prove to be incompatible with our knowledge, or necessary want of knowledge, some other hypostasis of men's hopes, genuine enough and worthy enough to replace it, will arise. But that the incongruous mixture of bad science with eviscerated papistry, out of which Comte manufactured the positivist religion, will be the heir of the Christian ages, I have too much respect for the humanity of the future to believe. Charles the Second told his brother, "They will not kill me, James, to make you king." And if critical science is remorselessly destroying the historical foundations of the noblest ideal of humanity which mankind have yet worshipped, it is little likely to permit the pitiful reality to climb into the vacant shrine.

That a man should determine to devote himself to the service of humanity—including intellectual and moral self-culture under that name; that this should be, in the proper sense of the word, his religion—is not only an intelligible, but, I think, a laudable resolution. And I am greatly disposed to believe that it is the only religion which will prove itself to be unassailably acceptable so long as the human race endures. But when the Comtist asks me to worship "Humanity"—that is to say, to adore the generalised conception of men as they ever have been and probably ever will be—I must reply that I could just as soon bow down and worship the generalised conception of a "wilderness of apes." Surely we are not going back to the days of Paganism, when individual men were deified, and the hard good sense of a dying Vespasian

could prompt the bitter jest, "Ut puto Deus fio."
No divinity doth hedge a modern man, be he
even a sovereign ruler. Nor is there any one,
except a municipal magistrate, who is officially
declared worshipful. But if there is no spark of
worship-worthy divinity in the individual twigs
of humanity, whence comes that godlike splen-
dour which the Moses of Positivism fondly
imagines to pervade the whole bush?

I know no study which is so unutterably sad-
dening as that of the evolution of humanity, as it
is set forth in the annals of history. Out of the
darkness of prehistoric ages man emerges with the
marks of his lowly origin strong upon him. He
is a brute, only more intelligent than the other
brutes, a blind prey to impulses, which as often
as not lead him to destruction; a victim to
endless illusions, which make his mental existence
a terror and a burden, and fill his physical life
with barren toil and battle. He attains a certain
degree of physical comfort, and develops a more or
less workable theory of life, in such favourable
situations as the plains of Mesopotamia or of
Egypt, and then, for thousands and thousands of
years, struggles, with varying fortunes, attended by
infinite wickedness, bloodshed, and misery, to
maintain himself at this point against the greed
and the ambition of his fellow-men. He makes a
point of killing and otherwise persecuting all
those who first try to get him to move on; and

when he has moved on a step, foolishly confers post-mortem deification on his victims. He exactly repeats the process with all who want to move a step yet farther. And the best men of the best epochs are simply those who make the fewest blunders and commit the fewest sins.

That one should rejoice in the good man, forgive the bad man, and pity and help all men to the best of one's ability, is surely indisputable. It is the glory of Judaism and of Christianity to have proclaimed this truth, through all their aberrations. But the worship of a God who needs forgiveness and help, and deserves pity every hour of his existence, is no better than that of any other voluntarily selected fetish. The Emperor Julian's project was hopeful in comparison with the prospects of the Comtist Anthropolatry.

When the historian of religion in the twentieth century is writing about the nineteenth, I foresee he will say something of this kind :

The most curious and instructive events in the religious history of the preceding century are the rise and progress of two new sects called Mormons and Positivists. To the student who has carefully considered these remarkable phenomena nothing in the records of religious self-delusion can appear improbable.

The Mormons arose in the midst of the great

Republic, which, though comparatively insignificant, at that time, in territory as in the number of
its citizens, was (as we know from the fragments
of the speeches of its orators which have come
down to us) no less remarkable for the native
intelligence of its population than for the wide
extent of their information, owing to the activity
of their publishers in diffusing all that they could
invent, beg, borrow, or steal. Nor were they less
noted for their perfect freedom from all restraints
in thought, or speech, or deed ; except, to be sure,
the beneficent and wise influence of the majority,
exerted, in case of need, through an institution
known as "tarring and feathering," the exact
nature of which is now disputed.

There is a complete consensus of testimony that
the founder of Mormonism, one Joseph Smith, was
a low-minded, ignorant scamp, and that he stole
the "Scriptures" which he propounded ; not being
clever enough to forge even such contemptible stuff
as they contain. Nevertheless he must have been
a man of some force of character, for a considerable
number of disciples soon gathered about him. In
spite of repeated outbursts of popular hatred and
violence—during one of which persecutions Smith
was brutally murdered—the Mormon body steadily
increased, and became a flourishing community.
But the Mormon practices being objectionable to
the majority, they were, more than once, without
any pretence of law, but by force of riot, arson, and

murder, driven away from the land they had occupied. Harried by these persecutions, the Mormon body eventually committed itself to the tender mercies of a desert as barren as that of Sinai; and after terrible sufferings and privations, reached the Oasis of Utah. Here it grew and flourished, sending out missionaries to, and receiving converts from, all parts of Europe, sometimes to the number of 10,000 in a year; until, in 1880, the rich and flourishing community numbered 110,000 souls in Utah alone, while there were probably 30,000 or 40,000 scattered abroad elsewhere. In the whole history of religions there is no more remarkable example of the power of faith; and, in this case, the founder of that faith was indubitably a most despicable creature. It is interesting to observe that the course taken by the great Republic and its citizens runs exactly parallel with that taken by the Roman Empire and its citizens towards the early Christians, except that the Romans had a certain legal excuse for their acts of violence, inasmuch as the Christian "sodalitia" were not licensed, and consequently were, *ipso facto*, illegal assemblages. Until, in the latter part of the nineteenth century, the United States legislature decreed the illegality of polygamy, the Mormons were wholly within the law.

Nothing can present a greater contrast to all this than the history of the Positivists. This sect arose much about the same time as that of the

Well, I will put your shell over it, and so, as
schoolboys make a spectre out of a turnip and a
tallow candle, behold the new religion of Humanity
complete !"

Unfortunately neither the Romanists, nor the
people who were something more than amateurs
in science, could be got to worship M. Comte's
new idol properly. In the native country of
Positivism, one distinguished man of letters and
one of science, for a time, helped to make up a
roomful. of the faithful, but their love soon grew
cold. In England, on the other hand, there appears to be little doubt that, in the ninth decade
of the century, the multitude of disciples reached
the grand total of several score. They had the
advantage of the advocacy of one or two most
eloquent and learned apostles, and, at any rate,
the sympathy of several persons of light and
leading; and, if they were not seen, they were
heard, all over the world. On the other hand, as
a sect, they laboured under the prodigious
disadvantage of being refined, estimable people,
living in the midst of the worn-out civilisation of
the old world; where any one who had tried to
persecute them, as the Mormons were persecuted,
would have been instantly hanged. But the
majority never dreamed of persecuting them; on
the contrary, they were rather given to scold and
otherwise try the patience of the majority.

The history of these sects in the closing years

of the century is highly instructive.   Mormonism . . . .

But I find I have suddenly slipped off Mr.
Harrison's tripod, which I had borrowed for the
occasion.   The fact is, I am not equal to the
prophetical business, and ought not to have
undertaken it.

[It did not occur to me, while writing the
latter part of this essay, that it could be needful
to disclaim the intention of putting the religious
system of Comte on a level with Mormonism.
And I was unaware of the fact that Mr. Harrison
rejects the greater part of the Positivist Religion,
as taught by Comte.   I have, therefore, erased
one or two passages, which implied his adherence
to the " Religion of Humanity " as developed by
Comte, 1893.]

VIII

AGNOSTICISM: A REJOINDER

[1889.]

THOSE who passed from Dr. Wace's article in the last number of the "Nineteenth Century" to the anticipatory confutation of it which followed in "The New Reformation," must have enjoyed the pleasure of a dramatic surprise—just as when the fifth act of a new play proves unexpectedly bright and interesting. Mrs. Ward will, I hope, pardon the comparison, if I say that her effective clearing away of antiquated incumbrances from the lists of the controversy, reminds me of nothing so much as of the action of some neat-handed, but strong-wristed, Phyllis, who, gracefully wielding her long-handled "Turk's head," sweeps away the accumulated results of the toil of generations of spiders. I am the more indebted to this luminous sketch of the results of critical investigation, as it is carried out among those theologians who are men of science and not mere counsel for creeds, since it has relieved me from the necessity of dealing with the greater part of Dr. Wace's polemic, and enables me to devote more space to the really important issues which have been raised.[1]

Perhaps, however, it may be well for me to observe that approbation of the manner in which a great biblical scholar, for instance, Reuss, does his work does not commit me to the adoption of all, or indeed any of his views ; and, further, that the disagreements of a series of investigators do not in any way interfere with the fact that each of them has made important contributions to the body of truth ultimately established.   If I cite Buffon, Linnæus, Lamarck, and Cuvier, as having each and all taken a leading share in building up modern biology, the statement that every one of these great naturalists disagreed with, and even more or less contradicted, all the rest is quite true ; but the supposition that the latter assertion is in any way inconsistent with the former, would betray a strange ignorance of the manner in which all true science advances.

Dr. Wace takes a great deal of trouble to make it appear that I have desired to evade the real questions raised by his attack upon me at the

---

[1] I may perhaps return to the question of the authorship of the Gospels. For the present I must content myself with warning my readers against any reliance upon Dr. Wace's statements as to the results arrived at by modern criticism. They are as gravely as surprisingly erroneous.

Church Congress. I assure the reverend Principal
that in this, as in some other respects, he has
entertained a very erroneous conception of my
intentions. Things would assume more accurate
proportions in Dr. Wace's mind, if he would
kindly remember that it is just thirty years since
ecclesiastical thunderbolts began to fly about my
ears. I have had the "Lion and the Bear" to
deal with, and it is long since I got quite used to
the threatenings of episcopal Goliaths, whose
croziers were like unto a weaver's beam. So that
I almost think I might not have noticed Dr.
Wace's attack, personal as it was ; and although,
as he is good enough to tell us, separate copies
are to be had for the modest equivalent of twopence,
as a matter of fact, it did not come under my
notice for a long time after it was made. May I
further venture to point out that (reckoning post-
age) the expenditure of twopence-halfpenny, or, at
the most, threepence, would have enabled Dr.
Wace so far to comply with ordinary conventions,
as to direct my attention to the fact that he had
attacked me before a meeting at which I was not
present ? I really am not responsible for the five
months' neglect of which Dr. Wace complains.
Singularly enough, the Englishry who swarmed
about the Engadine, during the three months
that I was being brought back to life by the
glorious air and perfect comfort of the Maloja, did
not, in my hearing, say anything about the

ability, I have satisfied myself of the soundness of
the foundations on which my arguments are built,
and I desire to be held fully responsible for
everything I say. But, nevertheless, my position
is really no more than that of an expositor; and
my justification for undertaking it is simply that
conviction of the supremacy of private judgment
(indeed, of the impossibility of escaping it) which
is the foundation of the Protestant Reformation,
and which was the doctrine accepted by the vast
majority of the Anglicans of my youth, before
that backsliding towards the " beggarly rudiments " of an effete and idolatrous sacerdotalism
which has, even now, provided us with the saddest
spectacle which has been offered to the eyes of
Englishmen in this generation. A high court of
ecclesiastical jurisdiction, with a host of great
lawyers in battle array, is and, for Heaven knows
how long, will be, occupied with these very
questions of " washing of cups and pots and brazen
vessels," which the Master, whose professed

truth, I think men of common sense would go elsewhere to learn
astronomy. Zeller's *Vorträge und Abhandlungen* were published
and came into my hands a quarter of a century ago. The
writer's rank, as a theologian to begin with, and subsequently
as a historian of Greek philosophy, is of the highest. Among
these essays are two—*Das Urchristenthum* and *Die Tübinger
historische Schule*—which are likely to be of more use to those
who wish to know the real state of the case than all that the
official "apologists," with their one eye on truth and the other
on the tenets of their sect, have written. For the opinion of a
scientific theologian about theologians of this stamp see pp. 225
and 227 of the *Vorträge.*

representatives are rending the Church over these
squabbles, had in his mind when, as we are told,
he uttered the scathing rebuke :—

Well did Isaiah prophesy of you hypocrites, as it is written,
    This people honoureth me with their lips,
    But their heart is far from me.
    But in vain do they worship me,
    Teaching as their doctrines the precepts of men.
                                        (Mark vii. 6-7.)

Men who can be absorbed in bickerings over
miserable disputes of this kind can have but little
sympathy with the old evangelical doctrine of the
"open Bible," or anything but a grave misgiving
of the results of diligent reading of the Bible,
without the help of ecclesiastical spectacles, by
the mass of the people. Greatly to the surprise
of many of my friends, I have always advocated
the reading of the Bible, and the diffusion of the
study of that most remarkable collection of books
among the people. Its teachings are so infinitely
superior to those of the sects, who are just as busy
now as the Pharisees were eighteen hundred years
ago, in smothering them under "the precepts of
men"; it is so certain, to my mind, that the Bible
contains within itself the refutation of nine-tenths
of the mixture of sophistical metaphysics and
old-world superstition which has been piled round
it by the so-called Christians of later times; it is
so clear that the only immediate and ready
antidote to the poison which has been mixed with

Christianity, to the intoxication and delusion of
mankind, lies in copious draughts from the
undefiled spring, that I exercise the right and
duty of free judgment on the part of every man,
mainly for the purpose of inducing other laymen
to follow my example.   If the New Testament
is translated into Zulu by Protestant missionaries,
it must be assumed that a Zulu convert is competent to draw from its contents all the truths which
it is necessary for him to believe.   I trust that I
may, without immodesty, claim to be put on the
same footing as a Zulu.

The most constant reproach which is launched
against persons of my way of thinking is that it is
all very well for us to talk about the deductions
of scientific thought, but what are the poor and
the uneducated to do ?   Has it ever occurred to
those who talk in this fashion, that their creeds
and the articles of their several confessions, their
determination of the exact nature and extent of
the teachings of Jesus, their expositions of the
real meaning of that which is written in the
Epistles (to leave aside all questions concerning
the Old Testament), are nothing more than
deductions which, at any rate, profess to be the
result of strictly scientific thinking, and which are
not worth attending to unless they really possess
that character ?   If it is not historically true that
such and such things happened in Palestine
eighteen centuries ago, what becomes of Chris-

tianity ?   And what is historical truth but that of
which the evidence bears strict scientific investigation ?   I do not call to mind any problem of
natural science which has come under my notice
which is more difficult, or more curiously
interesting as a mere problem, than that of the
origin of the Synoptic Gospels and that of the
historical value of the narratives which they
contain.   The Christianity of the Churches stands
or falls by the results of the purely scientific
investigation of these questions.   They were first
taken up, in a purely scientific spirit, about a
century ago ; they have been studied over and
over again by men of vast knowledge and critical
acumen ; but he would be a rash man who should
assert that any solution of these problems, as yet
formulated, is exhaustive.   The most that can be
said is that certain prevalent solutions are
certainly false, while others are more or less
probably true.

If I am doing my best to rouse my countrymen
out of their dogmatic slumbers, it is not that they
may be amused by seeing who gets the best of it
in a contest between a " scientist " and a theologian.   The serious question is whether theological
men of science, or theological special pleaders, are
to have the confidence of the general public;
it is the question whether a country in which it is
possible for a body of excellent clerical and lay
gentlemen to discuss, in public meeting assembled,

how much it is desirable to let the congregations
of the faithful know of the results of biblical
criticism, is likely to wake up with anything short
of the grasp of a rough lay hand upon its
shoulder; it is the question whether the New
Testament books, being, as I believe they were,
written and compiled by people who, according to
their lights, were perfectly sincere, will not, when
properly studied as ordinary historical documents,
afford us the means of self-criticism.   And it must
be remembered that the New Testament books
are not responsible for the doctrine invented by
the Churches that they are anything but ordinary
historical documents.   The author of the third
gospel tells us, as straightforwardly as a man can,
that he has no claim to any other character than
that of an ordinary compiler and editor, who had
before him the works of many and variously
qualified predecessors.

In my former papers, according to Dr. Wace, I
have evaded giving an answer to his main proposition, which he states as follows—

Apart from all disputed points of criticism, no one practically
doubts that our Lord lived, and that He died on the cross, in
the most intense sense of filial relation to His Father in Heaven,
and that He bore testimony to that Father's providence, love,
and grace towards mankind. The Lord's Prayer affords a
sufficient evidence on these points. If the Sermon on the Mount
alone be added, the whole unseen world, of which the Agnostic
refuses to know anything, stands unveiled before us. . . . If

Jesus Christ preached that Sermon, made those promises, and taught that prayer, then any one who says that we know nothing of God, or of a future life, or of an unseen world, says that he does not believe Jesus Christ (pp. 354-355).

Again—

> The main question at issue, in a word, is one which Professor Huxley has chosen to leave entirely on one side—whether, namely, allowing for the utmost uncertainty on other points of the criticism to which he appeals, there is any reasonable doubt that the Lord's Prayer and the Sermon on the Mount afford a true account of our Lord's essential belief and cardinal teaching (p. 355).

I certainly was not aware that I had evaded the questions here stated ; indeed I should say that I have indicated my reply to them pretty clearly; but, as Dr. Wace wants a plainer answer, he shall certainly be gratified. If, as Dr. Wace declares it is, his " whole case is involved in " the argument as stated in the latter of these two extracts, so much the worse for his whole case. For I am of opinion that there is the gravest reason for doubting whether the " Sermon on the Mount " was ever preached, and whether the so-called " Lord's Prayer " was ever prayed, by Jesus of Nazareth. My reasons for this opinion are, among others, these :—There is now no doubt that the three Synoptic Gospels, so far from being the work of three independent writers, are closely inter-dependent,[1] and that in one of two ways. Either

---

[1] I suppose this is what Dr. Wace is thinking about when he says that I allege that there "is no visible escape " from the

all three contain, as their foundation, versions, to
a large extent verbally identical, of one and the
same tradition ; or two of them are thus closely
dependent on the third ; and the opinion of the
majority of the best critics has of late years more
and more converged towards the conviction that
our canonical second gospel (the so-called "Mark's"
Gospel) is that which most closely represents the
primitive groundwork of the three.[1]   That I take
to be one of the most valuable results of New
Testament criticism, of immeasurably greater importance than the discussion about dates and
authorship.

But if, as I believe to be the case, beyond any
rational doubt or dispute, the second gospel is the
nearest extant representative of the oldest tradition, whether written or oral, how comes it that it

supposition of an *Ur-Marcus* (p. 367).   That a "theologian of
repute" should confound an indisputable fact with one of the
modes of explaining that fact is not so singular as those who are
unaccustomed to the ways of theologians might imagine.

[1] Any examiner whose duty it has been to examine into a case
of "copying" will be particularly well prepared to appreciate
the force of the case stated in that most excellent little book,
The Common Tradition of the Synoptic Gospels, by Dr. Abbott
and Mr. Rushbrooke (Macmillan, 1884).   To those who have not
passed through such painful experiences I may recommend the
brief discussion of the genuineness of the "Casket Letters" in my
friend Mr. Skelton's interesting book, *Maitland of Lethington*.
The second edition of Holtzmann's *Lehrbuch*, published in 1886,
gives a remarkably fair and full account of the present results of
criticism.   At p. 366 he writes that the present burning question
is whether the "relatively primitive narrative and the root of
the other synoptic texts is contained in Matthew or in Mark.
It is only on this point that properly-informed (*sachkundige*)
critics differ," and he decides in favour of Mark.

contains neither the " Sermon on the Mount " nor
the " Lord's Prayer," those typical embodiments,
according to Dr. Wace, of the " essential belief and
cardinal teaching" of Jesus ? Not only does
" Mark's " gospel fail to contain the " Sermon on
the Mount," or anything but a very few of the
sayings contained in that collection; but, at the
point of the history of Jesus where the " Sermon "
occurs in " Matthew," there is in " Mark " an
apparently unbroken narrative from the calling of
James and John to the healing of Simon's wife's
mother.   Thus the oldest tradition not only ignores
the " Sermon on the Mount," but, by implication,
raises a probability against its being delivered
when and where the later " Matthew " inserts it in
his compilation.

And still more weighty is the fact that the third
gospel, the author of which tells us that he wrote
after " many " others had " taken in hand " the
same enterprise; who should therefore have known
the first gospel (if it existed), and was bound to
pay to it the deference due to the work of an
apostolic eye-witness (if he had any reason for
thinking it was so)—this writer, who exhibits far
more literary competence than the other two,
ignores any " Sermon on the Mount," such as that
reported by " Matthew," just as much as the oldest
authority does. Yet " Luke " has a great many
passages identical, or parallel, with those in
" Matthew's " " Sermon on the Mount," which are,

for the most part, scattered about in a totally different connection.

Interposed, however, between the nomination of the Apostles and a visit to Capernaum ; occupying, therefore, a place which answers to that of the " Sermon on the Mount," in the first gospel, there is, in the third gospel a discourse which is as closely similar to the " Sermon in the Mount," in some particulars, as it is widely unlike it in others.

This discourse is said to have been delivered in a " plain " or " level place " (Luke vi. 17), and by way of distinction we may call it the " Sermon on the Plain."

I see no reason to doubt that the two Evangelists are dealing, to a considerable extent, with the same traditional material ; and a comparison of the two " Sermons " suggests very strongly that " Luke's " version is the earlier. The corresponddences between the two forbid the notion that they are independent. They both begin with a series of blessings, some of which are almost verbally identical. In the middle of each (Luke vi. 27-38, Matt. v. 43-48) there is a striking exposition of the ethical spirit of the command given in Leviticus xix. 18. And each ends with a passage containing the declaration that a tree is to be known by its fruit, and the parable of the house built on the sand. But while there are only 29 verses in the " Sermon on the Plain " there are 107 in the " Sermon on the Mount " ; the excess in length of the latter being chiefly due to the

long interpolations, one of 30 verses before and
one of 34 verses after, the middlemost parallelism
with Luke.   Under these circumstances it is quite
impossible to admit that there is more probability
that " Matthew's " version of the Sermon is historically accurate, than there is that Luke's version is
so ; and they cannot both be accurate.

" Luke " either knew the collection of loosely-
connected and aphoristic utterances which appear
under the name of the " Sermon on the Mount"
in " Matthew " ; or he did not.   If he did not, he
must have been ignorant of the existence of such
a document as our canonical " Matthew," a fact
which does not make for the genuineness, or the
authority, of that book.   If he did, he has shown
that he does not care for its authority on a matter
of fact of no small importance ; and that does not
permit us to conceive that he believed the first gospel
to be the work of an authority to whom he ought
to defer, let alone that of an apostolic eye-
witness.

The tradition of the Church about the second
gospel, which I believe to be quite worthless, but
which is all the evidence there is for " Mark's "
authorship, would have us believe that "Mark"
was little more than the mouthpiece of the apostle
Peter.   Consequently, we are to suppose that
Peter either did not know, or did not care very
much for, that account of the "essential belief
and cardinal teaching" of Jesus which is contained in the Sermon on the Mount ; and, certainly,

he could not have shared Dr. Wace's view of its importance.[1]

I thought that all fairly attentive and intelligent students of the gospels, to say nothing of theologians of reputation, knew these things. But how can any one who does know them have the conscience to ask whether there is " any reasonable doubt " that the Sermon on the Mount was preached by Jesus of Nazareth ? If conjecture is permissible, where nothing else is possible, the most probable conjecture seems to be that " Matthew," having a *cento* of sayings attributed— rightly or wrongly it is impossible to say—to Jesus among his materials, thought they were, or might be, records of a continuous discourse, and put them in at the place he thought likeliest.   Ancient historians of the highest character saw no harm in composing long speeches which never were spoken, and putting them into the mouths of statesmen and warriors ; and I presume that whoever is represented by " Matthew " would have been grievously astonished to find that any one objected to his following the example of the best models accessible to him.

[1] Holtzmann (*Die synoptischen Evangelien*, 1863, p. 75), following Ewald, argues that the ' Source A " (=the threefold tradition, more or less) contained something that answered to the "Sermon on the Plain " immediately after the words of our present Mark, "And he cometh into a house" (iii. 19).   But what conceivable motive could "Mark" have for omitting it ? Holtzmann has no doubt, however, that the "Sermon on the Mount" is a compilation, or, as he calls it in his recently-published *Lehrbuch* (p. 372), "an artificial mosaic work."

So with the "Lord's Prayer." Absent in our representative of the oldest tradition, it appears in both "Matthew" and "Luke." There is reason to believe that every pious Jew, at the commencement of our era, prayed three times a day, according to a formula which is embodied in the present "Schmone-Esre"[1] of the Jewish prayer-book. Jesus, who was assuredly, in all respects, a pious Jew, whatever else he may have been, doubtless did the same. Whether he modified the current formula, or whether the so-called "Lord's Prayer" is the prayer substituted for the "Schmone-Esre" in the congregations of the Gentiles, is a question which can hardly be answered.

In a subsequent passage of Dr. Wace's article (p. 356) he adds to the list of the verities which he imagines to be unassailable, "The Story of the Passion." I am not quite sure what he means by this. I am not aware that any one (with the exception of certain ancient heretics) has propounded doubts as to the reality of the crucifixion; and certainly I have no inclination to argue about the precise accuracy of every detail of that pathetic story of suffering and wrong. But, if Dr. Wace means, as I suppose he does, that that which, according to the orthodox view, happened after the crucifixion, and which is, in a dogmatic sense, the most important part of the story, is

---

[1] See Schürer, *Geschichte des jüdischen Volkes*, Zweiter Theil, p. 384.

founded on solid historical proofs, I must beg leave
to express a diametrically opposite conviction.

What do we find when the accounts of the
events in question, contained in the three Synoptic
gospels, are compared together? In the oldest,
there is a simple, straightforward statement which,
for anything that I have to urge to the contrary,
may be exactly true. In the other two, there is,
round this possible and probable nucleus, a mass
of accretions of the most questionable character.

The cruelty of death by crucifixion depended
very much upon its lingering character. If there
were a support for the weight of the body, as not
unfrequently was the practice, the pain during
the first hours of the infliction was not, necessarily,
extreme ; nor need any serious physical symptoms,
at once, arise from the wounds made by the nails
in the hands and feet, supposing they were nailed,
which was not invariably the case. When
exhaustion set in, and hunger, thirst, and nervous
irritation had done their work, the agony of the
sufferer must have been terrible ; and the more
terrible that, in the absence of any effectual
disturbance of the machinery of physical life, it
might be prolonged for many hours, or even days.
Temperate, strong men, such as were the ordinary
Galilean peasants, might live for several days on
the cross. It is necessary to bear these facts in
mind when we read the account contained in the
fifteenth chapter of the second gospel.

Jesus was crucified at the third hour (xv. 25), and the narrative seems to imply that he died immediately after the ninth hour (*v.* 34). In this case, he would have been crucified only six hours; and the time spent on the cross cannot have been much longer, because Joseph of Arimathæa must have gone to Pilate, made his preparations, and deposited the body in the rock-cut tomb before sunset, which, at that time of the year, was about the twelfth hour. That any one should die after only six hours' crucifixion could not have been at all in accordance with Pilate's large experience of the effects of that method of punishment. It, therefore, quite agrees with what might be expected, that Pilate " marvelled if he were already dead " and required to be satisfied on this point by the testimony of the Roman officer who was in command of the execution party. Those who have paid attention to the extraordinarily difficult question, What are the indisputable signs of death ?—will be able to estimate the value of the opinion of a rough soldier on such a subject; even if his report to the Procurator were in no wise affected by the fact that the friend of Jesus, who anxiously awaited his answer, was a man of influence and of wealth.

The inanimate body, wrapped in linen, was deposited in a spacious,[1] cool rock chamber, the

---

[1] Spacious, because a young man could sit in it " on the right side " (xv. 5), and therefore with plenty of room to spare.

entrance of which was closed, not by a well-fitting
door, but by a stone rolled against the opening,
which would of course allow free passage of air.
A little more than thirty-six hours afterwards
(Friday 6 P.M., to Sunday 6 A.M., or a little after)
three women visit the tomb and find it empty.
And they are told by a young man "arrayed in a
white robe" that Jesus is gone to his native
country of Galilee, and that the disciples and Peter
will find him there.

Thus it stands, plainly recorded, in the oldest
tradition that, for any evidence to the contrary,
the sepulchre may have been emptied at any time
during the Friday or Saturday nights.   If it is
said that no Jew would have violated the Sabbath
by taking the former course, it is to be recollected
that Joseph of Arimathæa might well be familiar
with that wise and liberal interpretation of the
fourth commandment, which permitted works of
mercy to men—nay, even the drawing of an ox or
an ass out of a pit—on the Sabbath.   At any
rate, the Saturday night was free to the most
scrupulous of observers of the Law.

These are the facts of the case as stated by the
oldest extant narrative of them.   I do not see why
any one should have a word to say against the
inherent probability of that narrative ; and, for my
part, I am quite ready to accept it as an historical
fact, that so much and no more is positively known
of the end of Jesus of Nazareth.   On what

grounds can a reasonable man be asked to believe
any more ?   So far as the narrative in the first
gospel, on the one hand, and those in the third
gospel and the Acts, on the other, go beyond what
is stated in the second gospel, they are hopelessly
discrepant with one another.   And this is the more
significant because the pregnant phrase "some
doubted," in the first gospel, is ignored in the
third.

But it is said that we have the witness Paul
speaking to us directly in the Epistles.   There is
little doubt that we have, and a very singular
witness he is.   According to his own showing,
Paul, in the vigour of his manhood, with every
means of becoming acquainted, at first hand, with
the evidence of eye-witnesses, not merely refused
to credit them, but " persecuted the church of God
and made havoc of it."   The reasoning of Stephen
fell dead upon the acute intellect of this zealot for
the traditions of his fathers : his eyes were blind
to the ecstatic illumination of the martyr's
countenance " as it had been the face of an
angel ; " and when, at the words " Behold, I see
the heavens opened and the Son of Man standing
on the right hand of God," the murderous mob
rushed upon and stoned the rapt disciple of Jesus,
Paul ostentatiously made himself their official
accomplice.

Yet this strange man, because he has a vision
one day, at once, and with equally headlong zeal,

flies to the opposite pole of opinion. And he is most careful to tell us that he abstained from any re-examination of the facts.

Immediately I conferred not with flesh and blood ; neither went I up to Jerusalem to them which were Apostles before me ; but I went away into Arabia. (Galatians i. 16, 17.)

I do not presume to quarrel with Paul's procedure. If it satisfied him, that was his affair ; and, if it satisfies anyone else, I am not called upon to dispute the right of that person to be satisfied. But I certainly have the right to say that it would not satisfy me, in like case ; that I should be very much ashamed to pretend that it could, or ought to, satisfy me ; and that I can entertain but a very low estimate of the value of the evidence of people who are to be satisfied in this fashion, when questions of objective fact, in which their faith is interested, are concerned. So that when I am called upon to believe a great deal more than the oldest gospel tells me about the final events of the history of Jesus on the authority of Paul (1 Corinthians xv. 5-8) I must pause. Did he think it, at any subsequent time, worth while " to confer with flesh and blood," or, in modern phrase, to re-examine the facts for himself ? or was he ready to accept anything that fitted in with his preconceived ideas ? Does he mean, when he speaks of all the appearances of Jesus after the crucifixion as if they were of the same kind, that

they were all visions, like the manifestation to
himself ?   And, finally, how is this account to be
reconciled with those in the first and third
gospels—which, as we have seen, disagree with
one another ?

Until these questions are satisfactorily answered,
I am afraid that, so far as I am concerned, Paul's
testimony cannot be seriously regarded, except as
it may afford evidence of the state of traditional
opinion at the time at which he wrote, say
between 55 and 60 A.D. ; that is, more than
twenty years after the event ; a period much more
than sufficient for the development of any amount
of mythology about matters of which nothing was
really known.   A few years later, among the contemporaries and neighbours of the Jews, and, if
the most probable interpretation of the Apocalypse can be trusted, among the followers of Jesus
also, it was fully believed, in spite of all the
evidence to the contrary, that the Emperor Nero
was not really dead, but that he was hidden away
somewhere in the East, and would speedily come
again at the head of a great army, to be revenged
upon his enemies.[1]

Thus, I conceive that I have shown cause for
the opinion that Dr. Wace's challenge touching
the Sermon on the Mount, the Lord's Prayer, and

---

[1] King Herod had not the least difficulty in supposing the
resurrection of John the Baptist—" John, whom I beheaded,
he is risen " (Mark vi. 16).

the Passion was more valorous than discreet.
After all this discussion, I am still at the agnostic
point. Tell me, first, what Jesus can be proved
to have been, said, and done, and I will say
whether I believe him, or in him,[1] or not. As Dr.
Wace admits that I have dissipated his lingering
shade of unbelief about the bedevilment of the
Gadarene pigs, he might have done something to
help mine. Instead of that, he manifests a total
want of conception of the nature of the obstacles
which impede the conversion of his "infidels."

The truth I believe to be, that the difficulties
in the way of arriving at a sure conclusion as to
these matters, from the Sermon on the Mount,
the Lord's Prayer, or any other data offered by
the Synoptic gospels (and *à fortiori* from the
fourth gospel), are insuperable. Every one of
these records is coloured by the prepossessions of
those among whom the primitive traditions arose,
and of those by whom they were collected and
edited : and the difficulty of making allowance for
these prepossessions is enhanced by our ignorance
of the exact dates at which the documents were
first put together; of the extent to which they

[1] I am very sorry for the interpolated " in," because citation
ought to be accurate in small things as in great. But what
difference it makes whether one "believes Jesus" or "believes
in Jesus" much thought has not enabled me to discover. If
you "believe him" you must believe him to be what he professed to be—that is, "believe in him;" and if you "believe
in him" you must necessarily "believe him."

have been subsequently worked over ana interpolated ; and of the historical sense, or want of
sense, and the dogmatic tendencies of their
compilers and editors.   Let us see if there is any
other road which will take us into something
better than negation.

There is a widespread notion that the "primitive Church," while under the guidance of the
Apostles and their immediate successors, was a
sort of dogmatic dovecot, pervaded by the most
loving unity and doctrinal harmony.   Protestants,
especially, are fond of attributing to themselves
the merit of being nearer "the Church of the
Apostles" than their neighbours ; and they are
the less to be excused for their strange delusion
because they are great readers of the documents
which prove the exact contrary.   The fact is that,
in the course of the first three centuries of its
existence, the Church rapidly underwent a process
of evolution of the most remarkable character,
the final stage of which is far more different from
the first than Anglicanism is from Quakerism.
The key to the comprehension of the problem
of the origin of that which is now called
"Christianity," and its relation to Jesus of
Nazareth, lies here.   Nor can we arrive at any
sound conclusion as to what it is probable that
Jesus actually said and did, without being clear on
this head.   By far the most important and
subsequently influential steps in the evolution of

Christianity took place in the course of the century, more or less, which followed upon the crucifixion. It is almost the darkest period of Church history, but, most fortunately, the beginning and the end of the period are brightly illuminated by the contemporary evidence of two writers of whose historical existence there is no doubt,[1] and against the genuineness of whose most important works there is no widely-admitted objection. These are Justin, the philosopher and martyr, and Paul, the Apostle to the Gentiles. I shall call upon these witnesses only to testify to the condition of opinion among those who called themselves disciples of Jesus in their time.

Justin, in his Dialogue with Trypho the Jew, which was written somewhere about the middle of the second century, enumerates certain categories of persons who, in his opinion, will, or will not, be saved.[2] These are :—

1. Orthodox Jews who refuse to believe that Jesus is the Christ. *Not Saved.*

2. Jews who observe the Law; believe Jesus to be the Christ; but who insist on the observance of the Law by Gentile converts. *Not Saved.*

3. Jews who observe the Law; believe Jesus to

---

[1] True for Justin : but there is a school of theological critics, who more or less question the historical reality of Paul, and the genuineness of even the four cardinal epistles.

[2] See *Dial. cum Tryphone*, § 47 and § 35. It is to be understood that Justin does not arrange these categories in order, as I have done.

be the Christ, and hold that Gentile converts
need not observe the Law. *Saved* (in Justin's
opinion; but some of his fellow-Christians think
the contrary).

4. Gentile converts to the belief in Jesus as the
Christ, who observe the Law. *Saved* (possibly).

5. Gentile believers in Jesus as the Christ, who
do not observe the Law themselves (except so far
as the refusal of idol sacrifices), but do not
consider those who do observe it heretics. *Saved*
(this is Justin's own view).

6. Gentile believers who do not observe the
Law, except in refusing idol sacrifices, and hold
those who do observe it to be heretics. *Saved*.

7. Gentiles who believe Jesus to be the Christ
and call themselves Christians, but who eat meats
sacrificed to idols. *Not Saved*.

8. Gentiles who disbelieve in Jesus as the
Christ. *Not Saved*.

Justin does not consider Christians who believe
in the natural birth of Jesus, of whom he implies
that there is a respectable minority, to be heretics,
though he himself strongly holds the preternatural
birth of Jesus and his pre-existence as the
" Logos " or " Word." He conceives the Logos to
be a second God, inferior to the first, unknowable
God, with respect to whom Justin, like Philo, is
a complete agnostic. The Holy Spirit is not regarded by Justin as a separate personality, and
is often mixed up with the " Logos." The

doctrine of the natural immortality of the soul is,
for Justin, a heresy; and he is as firm a believer
in the resurrection of the body, as in the
speedy Second Coming and the establishment of
the millennium.

This pillar of the Church in the middle of
the second century—a much-travelled native of
Samaria—was certainly well acquainted with
Rome, probably with Alexandria; and it is likely
that he knew the state of opinion throughout the
length and breadth of the Christian world as well
as any man of his time.   If the various categories
above enumerated are arranged in a   series
thus :—

Justin's Christianity

<i>Orthodox Judaism</i>	<i>Judœo-Christianity</i>					<i>Idolothytic Christianity</i>	<i>Paganism</i>
I.	II.	III.	IV.	V.	VI.	VII.	VIII.

it is obvious that they form a gradational series
from orthodox Judaism, on the extreme left, to
Paganism, whether philosophic or popular, on the
extreme right; and it will further be observed
that, while Justin's conception of Christianity is
very broad, he rigorously excludes two classes of
persons who, in his time, called themselves
Christians; namely, those who insist on circumcision and other observances of the Law on the
part of Gentile converts; that is to say, the strict
Judæo-Christians (II.); and, on the other hand,
those who assert the lawfulness of eating meat

offered to idols—whether they are Gnostic or not
(VII.) These last I have called "idolothytic"
Christians, because I cannot devise a better
name, not because it is strictly defensible etymo-
logically.

At the present moment, I do not suppose there
is an English missionary in any heathen land who
would trouble himself whether the materials of his
dinner had been previously offered to idols or not.
On the other hand, I suppose there is no Protestant
sect within the pale of orthodoxy, to say nothing of
the Roman and Greek Churches, which would
hesitate to declare the practice of circumcision and
the observance of the Jewish Sabbath and dietary
rules, shockingly heretical.

Modern Christianity has, in fact, not only shifted
far to the right of Justin's position, but it is of
much narrower compass.

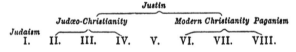

For, though it includes VII., and even, in saint
and relic worship, cuts a "monstrous cantle" out
of paganism, it excludes, not only all Judæo-
Christians, but all who doubt that such are
heretics. Ever since the thirteenth century, the
Inquisition would have cheerfully burned, and in
Spain did abundantly burn, all persons who came
under the categories II., III. IV., V. And the

wolf would play the same havoc now, if it could only get its blood-stained jaws free from the muzzle imposed by the secular arm.

Further, there is not a Protestant body except the Unitarian, which would not declare Justin himself a heretic, on account of his doctrine of the inferior godship of the Logos; while I am very much afraid that, in strict logic, Dr. Wace would be under the necessity, so painful to him, of calling him an "infidel," on the same and on other grounds.

Now let us turn to our other authority. If there is any result of critical investigations of the sources of Christianity which is certain,[1] it is that Paul of Tarsus wrote the Epistle to the Galatians somewhere between the years 55 and 60 A.D., that is to say, roughly, twenty, or five-and-twenty years after the crucifixion. If this is so, the Epistle to the Galatians is one of the oldest, if not the very oldest, of extant documentary evidences of the state of the primitive Church. And, be it observed, if it is Paul's writing, it unquestionably furnishes us with the evidence of a participator in the transactions narrated. With the exception of two or three of the other Pauline Epistles, there is not one solitary book in the New Testament of the authorship and authority of which we have such good evidence.

[1] I guard myself against being supposed to affirm that even the four cardinal epistles of Paul may not have been seriously tampered with. See note 1, p. 287 above.

And what is the state of things we find dis-
closed ?   A bitter quarrel, in his account of which
Paul by no means minces matters, or hesitates to
hurl defiant sarcasms against those who were
"reputed to be pillars" : James "the brother of
the Lord," Peter, the rock on whom Jesus is said
to have built his Church, and John, "the beloved
disciple."   And no deference toward "the rock"
withholds Paul from charging Peter to his face with
" dissimulation."

The subject of the hot dispute was simply this.
Were Gentile converts bound to obey the Law or
not ?   Paul answered in the negative ; and, acting
upon his opinion, he had created at Antioch (and
elsewhere) a specifically "Christian" community,
the sole qualifications for admission into which were
the confession of the belief that Jesus was the
Messiah, and baptism upon that confession.   In
the epistle in question, Paul puts this—his
"gospel," as he calls it—in its most extreme form.
Not only does he deny the necessity of conformity
with the Law, but he declares such conformity to
have a negative value.   "Behold, I, Paul, say
unto you, that if ye receive circumcision, Christ
will profit you nothing" (Galatians v. 2).   He
calls the legal observances " beggarly rudiments,"
and anathematises every one who preaches to the
Galatians any other gospel than his own.   That is
to say, by direct consequence, he anathematises the
Nazarenes of Jerusalem, whose zeal for the Law is

testified by James in a passage of the Acts cited further on. In the first Epistle to the Corinthians, dealing with the question of eating meat offered to idols, it is clear that Paul himself thinks it a matter of indifference; but he advises that it should not be done, for the sake of the weaker brethren. On the other hand, the Nazarenes of Jerusalem most strenuously opposed Paul's "gospel," insisting on every convert becoming a regular Jewish proselyte, and consequently on his observance of the whole Law; and this party was led by James and Peter and John (Galatians ii. 9) Paul does not suggest that the question of principle was settled by the discussion referred to in Galatians. All he says is, that it ended in the practical agreement that he and Barnabas should do as they had been doing, in respect to the Gentiles; while James and Peter and John should deal in their own fashion with Jewish converts. Afterwards, he complains bitterly of Peter, because, when on a visit to Antioch, he, at first, inclined to Paul's view and ate with the Gentile converts; but when " certain came from James," " drew back, and separated himself, fearing them that were of the circumcision. And the rest of the Jews dissembled likewise with him; insomuch that even Barnabas was carried away with their dissimulation" (Galatians ii. 12-13).

There is but one conclusion to be drawn from Paul's account of this famous dispute, the settle-

ment of which determined the fortunes of the
nascent religion. It is that the disciples at Jerusalem, headed by "James, the Lord's brother," and
by the leading apostles, Peter and John, were strict
Jews, who had objected to admit any converts
into their body, unless these, either by birth, or by
becoming proselytes, were also strict Jews. In
fact, the sole difference between James and Peter
and John, with the body of the disciples whom
they led and the Jews by whom they were
surrounded, and with whom they, for many years,
shared the religious observances of the Temple,
was that they believed that the Messiah, whom
the leaders of the nation yet looked for, had
already come in the person of Jesus of Nazareth.

The Acts of the Apostles is hardly a very trustworthy history; it is certainly of later date than
the Pauline Epistles, supposing them to be
genuine. And the writer's version of the conference of which Paul gives so graphic a description,
if that is correct, is unmistakably coloured with
all the art of a reconciler, anxious to cover up a
scandal. But it is none the less instructive on
this account. The judgment of the "council"
delivered by James is that the Gentile converts
shall merely "abstain from things sacrificed to
idols, and from blood and from things strangled,
and from fornication." But notwithstanding the
accommodation in which the writer of the Acts
would have us believe, the Jerusalem Church held

to its endeavour to retain the observance of the
Law. Long after the conference, some time after
the writing of the Epistles to the Galatians and
Corinthians, and immediately after the despatch of
that to the Romans, Paul makes his last visit to
Jerusalem, and presents himself to James and all
the elders. And this is what the Acts tells us of
the interview :—

> And they said unto him, Thou seest, brother, how many
> thousands [or myriads] there are among the Jews of them which
> have believed ; and they are all zealous for the law ; and they
> have been informed concerning thee, that thou teachest all the
> Jews which are among the Gentiles to forsake Moses, telling
> them not to circumcise their children, neither to walk after the
> customs. (Acts xxi. 20, 21.)

They therefore request that he should perform a
certain public religious act in the Temple, in
order that

> all shall know that there is no truth in the things whereof they
> have been informed concerning thee ; but that thou thyself
> walkest orderly, keeping the law (*ibid.* 24).[1]

How far Paul could do what he is here requested to do, and which the writer of the Acts
goes on to say he did, with a clear conscience, if he
wrote the Epistles to the Galatians and Corinthians, I may leave any candid reader of these
epistles to decide. The point to which I wish to

---

[1] [Paul, in fact, is required to commit in Jerusalem, an act
of the same character as that which he brands as "dissimulation" on the part of Peter in Antioch.]

direct attention is the declaration that the Jerusalem Church, led by the brother of Jesus and by his personal disciples and friends, twenty years and more after his death, consisted of strict and zealous Jews.

Tertullus, the orator, caring very little about the internal dissensions of the followers of Jesus, speaks of Paul as a "ringleader of the sect of the Nazarenes" (Acts xxiv. 5), which must have affected James much in the same way as it would have moved the Archbishop of Canterbury, in George Fox's day, to hear the latter called a "ringleader of the sect of Anglicans." In fact, "Nazarene" was, as is well known, the distinctive appellation applied to Jesus; his immediate followers were known as Nazarenes; while the congregation of the disciples, and, later, of converts at Jerusalem—the Jerusalem Church—was emphatically the "sect of the Nazarenes," no more, in itself, to be regarded as anything outside Judaism than the sect of the Sadducees, or that of the Essenes.[1]   In fact, the tenets of both the Sadducees and the Essenes diverged much more widely from the Pharisaic standard of orthodoxy than Nazarenism did.

Let us consider the position of affairs now (A.D. 50-60) in relation to that which obtained in

[1] All this was quite clearly pointed out by Ritschl nearly forty years ago. See *Die Entstehung der alt-katholischen Kirche* (1850), p. 108.

Justin's time, a century later. It is plain that
the Nazarenes—presided over by James, "the
brother of the Lord," and comprising within their
body all the twelve apostles—belonged to Justin's
second category of "Jews who observe the Law,
believe Jesus to be the Christ, but who insist on
the observance of the Law by Gentile converts,"
up till the time at which the controversy reported
by Paul arose. They then, according to Paul,
simply allowed him to form his congregations of
non-legal Gentile converts at Antioch and elsewhere; and it would seem that it was to these
converts, who would come under Justin's fifth
category, that the title of "Christian" was first
applied. If any of these Christians had acted
upon the more than half-permission given by
Paul, and had eaten meats offered to idols,
they would have belonged to Justin's seventh
category.

Hence, it appears that, if Justin's opinion,
which was probably that of the Church generally
in the middle of the second century, was correct,
James and Peter and John and their followers
could not be saved; neither could Paul, if he
carried into practice his views as to the indifference of eating meats offered to idols. Or, to put
the matter another way, the centre of gravity of
orthodoxy, which is at the extreme right of the
series in the nineteenth century, was at the extreme left, just before the middle of the first

century, when the "sect of the Nazarenes" constituted the whole church founded by Jesus and the
apostles; while, in the time of Justin, it lay mid-
way between the two.   It is therefore a profound
mistake to imagine that the Judæo-Christians
(Nazarenes and Ebionites) of later times were
heretical outgrowths from a primitive universalist
" Christianity."   On the contrary, the universalist
" Christianity "   is   an   outgrowth   from   the
primitive, purely Jewish, Nazarenism; which,
gradually eliminating all the ceremonial and
dietary parts of the Jewish law, has thrust aside
its parent, and all the intermediate stages of its
development, into the position of damnable
heresies.

Such being the case, we are in a position to
form a safe judgment of the limits within which
the teaching of Jesus of Nazareth must have been
confined.   Ecclesiastical authority would have us
believe that the words which are given at the end
of the first Gospel, " Go ye, therefore, and make
disciples of all the nations, baptizing them in the
name of the Father and of the Son and of the
Holy Ghost," are part of the last commands of
Jesus, issued at the moment of his parting with
the eleven.   If so, Peter and John must have
heard these words; they are too plain to be misunderstood; and the occasion is too solemn for
them ever to be forgotten.   Yet the " Acts " tells
us that Peter needed a vision to enable him so

much as to baptize Cornelius; and Paul, in the
Galatians, knows nothing of words which would
have completely borne him out as against those
who, though they heard, must be supposed to
have either forgotten, or ignored them. On the
other hand, Peter and John, who are supposed to
have heard the "Sermon on the Mount," know
nothing of the saying that Jesus had not come to
destroy the Law, but that every jot and tittle of
the Law must be fulfilled, which surely would
have been pretty good evidence for their view of
the question.

We are sometimes told that the personal
friends and daily companions of Jesus remained
zealous Jews and opposed Paul's innovations,
because they were hard of heart and dull of
comprehension. This hypothesis is hardly in
accordance with the concomitant faith of those
who adopt it, in the miraculous insight and super-
human sagacity of their Master ; nor do I see any
way of getting it to harmonise with the orthodox
postulate; namely, that Matthew was the author
of the first gospel and John of the fourth. If that
is so, then, most assuredly, Matthew was no
dullard; and as for the fourth gospel—a theo-
sophic romance of the first order—it could have
been written by none but a man of remarkable
literary capacity, who had drunk deep of
Alexandrian philosophy. Moreover, the doctrine
of the writer of the fourth gospel is more remote

from that of the " sect of the Nazarenes " than is
that of Paul himself. I am quite aware that
orthodox critics have been capable of maintaining
that John, the Nazarene, who was probably well
past fifty years of age, when he is supposed to have
written the most thoroughly Judaising book in
the New Testament—the Apocalypse—in the
roughest of Greek, underwent an astounding
metamorphosis of both doctrine and style by the
time he reached the ripe age of ninety or so, and
provided the world with a history in which the
acutest critic cannot [always] make out where the
speeches of Jesus end and the text of the narrative
begins; while that narrative is utterly irreconcilable, in regard to matters of fact, with that of his
fellow-apostle, Matthew.

The end of the whole matter is this :—The
" sect of the Nazarenes," the brother and the
immediate followers of Jesus, commissioned by
him as apostles, and those who were taught by
them up to the year 50 A.D., were not "Christians"
in the sense in which that term has been understood ever since its asserted origin at Antioch, but
Jews—strict orthodox Jews—whose belief in the
Messiahship of Jesus never led to their exclusion
from the Temple services, nor would have shut
them out from the wide embrace of Judaism.[1]

---

[1] " If every one was baptized as soon as he acknowledged Jesus
to be the Messiah, the first Christians can have been aware of no
other essential differences from the Jews."—Zeller, *Vorträge*
(1865), p. 26.

The open proclamation of their special view about
the Messiah was doubtless offensive to the
Pharisees, just as rampant Low Churchism is
offensive to bigoted High Churchism in our own
country ; or as any kind of dissent is offensive to
fervid religionists of all creeds. To the Sadducees,
no doubt, the political danger of any Messianic
movement was serious ; and they would have been
glad to put down Nazarenism, lest it should end
in useless rebellion against their Roman masters,
like that other Galilean movement headed by
Judas, a generation earlier. Galilee was always a
hotbed of seditious enthusiasm against the rule of
Rome ; and high priest and procurator alike had
need to keep a sharp eye upon natives of that
district. On the whole, however, the Nazarenes
were but little troubled for the first twenty years
of their existence ; and the undying hatred of the
Jews against those later converts, whom they
regarded as apostates and fautors of a sham
Judaism, was awakened by Paul. From their
point of view, he was a mere renegade Jew,
opposed alike to orthodox Judaism and to orthodox Nazarenism ; and whose teachings threatened
Judaism with destruction. And, from their point
of view, they were quite right. In the course of
a century, Pauline influences had a large share in
driving primitive Nazarenism from being the very
heart of the new faith into the position of scouted
error ; and the spirit of Paul's doctrine continued

Judaism. Again, who is to say whether Jesus proclaimed himself the veritable Messiah, expected by his nation since the appearance of the pseudo-prophetic work of Daniel, a century and a half before his time; or whether the enthusiasm of his followers gradually forced him to assume that position?

But one thing is quite certain: if that belief in the speedy second coming of the Messiah which was shared by all parties in the primitive Church, whether Nazarene or Pauline; which Jesus is made to prophesy, over and over again, in the Synoptic gospels; and which dominated the life of Christians during the first century after the crucifixion;—if he believed and taught that, then assuredly he was under an illusion, and he is responsible for that which the mere effluxion of time has demonstrated to be a prodigious error.

When I ventured to doubt "whether any Protestant theologian who has a reputation to lose will say that he believes the Gadarene story," it appears that I reckoned without Dr. Wace, who, referring to this passage in my paper, says :—

He will judge whether I fall under his description; but I repeat that I believe it, and that he has removed the only objection to my believing it (p. 363).

Far be it from me to set myself up as a judge

of any such delicate question as that put before
me ; but I think I may venture to express the
conviction that, in the matter of courage, Dr.
Wace has raised for himself a monument *ære
perennius.* For really, in my poor judgment, a
certain splendid intrepidity, such as one admires
in the leader of a forlorn hope, is manifested
by Dr. Wace when he solemnly affirms that he
believes the Gadarene story on the evidence
offered.   I feel less complimented perhaps than I
ought to do, when I am told that I have been an
accomplice in extinguishing in Dr. Wace's mind
the last glimmer of doubt which common sense
may have suggested.   In fact, I must disclaim all
responsibility for the use to which the information
I supplied has been put.   I formally decline to
admit that the expression of my ignorance whether
devils, in the existence of which I do not believe,
if they did exist, might or might not be made to
go out of men into pigs, can, as a matter of logic,
have been of any use whatever to a person who
already believed in devils and in the historical
accuracy of the gospels.

Of the Gadarene story, Dr. Wace, with all
solemnity and twice over, affirms that he " believes
it." I am sorry to trouble him further, but what
does he mean by " it " ?   Because there are two
stories, one in " Mark " and " Luke," and the other
in " Matthew." In the former, which I quoted
in my previous paper, there is one possessed

man; in the latter there are two. The story is
told fully, with the vigorous homely diction and
the picturesque details of a piece of folklore, in
the second gospel. The immediately antecedent
event is the storm on the Lake of Gennesaret.
The immediately consequent events are the
message from the ruler of the synagogue and the
healing of the woman with an issue of blood.
In the third gospel, the order of events is exactly
the same, and there is an extremely close general
and verbal correspondence between the narratives
of the miracle. Both agree in stating that there
was only one possessed man, and that he was
the residence of many devils, whose name was
" Legion."

In the first gospel, the event which immediately
precedes the Gadarene affair is, as before, the
storm; the message from the ruler and the healing
of the issue are separated from it by the accounts
of the healing of a paralytic, of the calling of
Matthew, and of a discussion with some Pharisees.
Again, while the second gospel speaks of the
country of the " Gerasenes " as the locality of the
event, the third gospel has " Gerasenes,"
" Gergesenes," and " Gadarenes " in different
ancient MSS. ; while the first has " Gadarenes."

The really important points to be noticed,
however, in the narrative of the first gospel, are
these—that there are two possessed men instead
of one; and that while the story is abbreviated by

omissions, what there is of it is often verbally
identical with the corresponding passages in the
other two gospels. The most unabashed of
reconcilers cannot well say that one man is the
same as two, or two as one; and, though the
suggestion really has been made, that two different
miracles, agreeing in all essential particulars,
except the number of the possessed, were effected
immediately after the storm on the lake, I should
be sorry to accuse any one of seriously adopting it.
Nor will it be pretended that the allegory refuge
is accessible in this particular case.

So, when Dr. Wace says that he believes in the
synoptic evangelists' account of the miraculous
bedevilment of swine, I may fairly ask which of
them does he believe? Does he hold by the one
evangelist's story, or by that of the two evangelists? And having made his election, what
reasons has he to give for his choice? If it is
suggested that the witness of two is to be taken
against that of one, not only is the testimony
dealt with in that common-sense fashion against
which the theologians of his school protest so
warmly.; not only is all question of inspiration at
an end, but the further inquiry arises, After all, is
it the testimony of two against one? Are the
authors of the versions in the second and third
gospels really independent witnesses? In order to
answer this question, it is only needful to place
the English versions of the two side by side, and

compare them carefully. It will then be seen that the coincidences between them, not merely in substance, but in arrangement, and in the use of identical words in the same order, are such, that only two alternatives are conceivable: either one evangelist freely copied from the other, or both based themselves upon a common source, which may either have been a written document, or a definite oral tradition learned by heart. Assuredly, these two testimonies are not those of independent witnesses. Further, when the narrative in the first gospel is compared with that in the other two, the same fact comes out.

Supposing, then, that Dr. Wace is right in his assumption that Matthew, Mark, and Luke wrote the works which we find attributed to them by tradition, what is the value of their agreement, even that something more or less like this particular miracle occurred, since it is demonstrable, either that all depend on some antecedent statement, of the authorship of which nothing is known, or that two are dependent upon the third?

Dr. Wace says he believes the Gadarene story; whichever version of it he accepts, therefore, he believes that Jesus said what he is stated in all the versions to have said, and thereby virtually declared that the theory of the nature of the spiritual world involved in the story is true. Now I hold that this theory is false, that it is a monstrous and mischievous fiction; and I unhesi-

tatingly express my disbelief in any assertion that
it is true, by whomsoever made.   So that, if Dr.
Wace is right in his belief, he is also quite right
in classing me among the people he calls "infidels";
and although I cannot fulfil the eccentric expectation that I shall glory in a title which, from my
point of view, it would be simply silly to adopt,
I certainly shall rejoice not to be reckoned among
"Christians" so long as the profession of belief in
such stories as the Gadarene pig affair, on the
strength of a tradition of unknown origin, of which
two discrepant reports, also of unknown origin,
alone remain, forms any part of the Christian
faith.   And, although I have, more than once,
repudiated the gift of prophecy, yet I think I
may venture to express the anticipation, that if
"Christians" generally are going to follow the line
taken by Dr. Wace, it will not be long before all
men of common sense qualify for a place among
the "infidels."

IX

AGNOSTICISM AND CHRISTIANITY

[1889]

Nemo ergo ex me scire quærat, quod me nescire scio, nisi forte ut nescire discat.—AUGUSTINUS, *De Civ. Dei*, xii. 7.

[1] THE present discussion has arisen out of the use, which has become general in the last few years, of the terms "Agnostic" and "Agnosticism."

The people who call themselves "Agnostics" have been charged with doing so because they have not the courage to declare themselves "Infidels." It has been insinuated that they have adopted a new name in order to escape the unpleasantness which attaches to their proper denomination. To this wholly erroneous imputation, I have replied by showing that the term "Agnostic" did, as a matter of fact, arise in a manner which negatives it; and my statement has not been, and cannot be, refuted. Moreover,

[1] The substance of a paragraph which precedes this has been transferred to the Prologue.

speaking for myself, and without impugning the right of any other person to use the term in another sense, I further say that Agnosticism is not properly described as a "negative" creed, nor indeed as a creed of any kind, except in so far as it expresses absolute faith in the validity of a principle, which is as much ethical as intellectual. This principle may be stated in various ways, but they all amount to this: that it is wrong for a man to say that he is certain of the objective truth of any proposition unless he can produce evidence which logically justifies that certainty. This is what Agnosticism asserts; and, in my opinion, it is all that is essential to Agnosticism. That which Agnostics deny and repudiate, as immoral, is the contrary doctrine, that there are propositions which men ought to believe, without logically satisfactory evidence; and that reprobation ought to attach to the profession of disbelief in such inadequately supported propositions.   The justification of the Agnostic principle lies in the success which follows upon its application, whether in the field of natural, or in that of civil, history; and in the fact that, so far as these topics are concerned, no sane man thinks of denying its validity.

Still speaking for myself, I add, that though Agnosticism is not, and cannot be, a creed, except in so far as its general principle is concerned; yet that the application of that principle results in the denial of, or the suspension of judgment concerning, a number of propositions respecting which our contemporary ecclesiastical "gnostics" profess entire certainty. And, in so far as these ecclesiastical persons can be justified in their old-established custom (which many nowadays think more honoured in the breach than the observance) of using opprobrious names to those who differ from them, I fully admit their right to call me and those who think with me "Infidels"; all I have ventured to urge is that they must not expect us to speak of ourselves by that title.

The extent of the region of the uncertain, the number of the problems the investigation of which ends in a verdict of not proven, will vary according to the knowledge and the intellectual habits of the individual Agnostic. I do not very much care to speak of anything as "unknowable." [1] What I am sure about is that there are many topics about which I know nothing; and which, so far as I can see, are out of reach of my faculties. But whether these things are knowable by any one else is exactly one of those matters which is beyond my knowledge, though I may have a tolerably strong opinion as to the probabilities of the case. Relatively to myself, I am quite sure that the region of uncertainty—the nebulous country in which words play the part of realities

[1] I confess that, long ago, I once or twice made this mistake ; even to the waste of a capital 'U.'   1893.

—is far more extensive than I could wish. Materialism and Idealism; Theism and Atheism; the doctrine of the soul and its mortality or immortality—appear in the history of philosophy like the shades of Scandinavian heroes, eternally slaying one another and eternally coming to life again in a metaphysical "Nifelheim." It is getting on for twenty-five centuries, at least, since mankind began seriously to give their minds to these topics. Generation after generation, philosophy has been doomed to roll the stone uphill; and, just as all the world swore it was at the top, down it has rolled to the bottom again. All this is written in innumerable books; and he who will toil through them will discover that the stone is just where it was when the work began. Hume saw this; Kant saw it; since their time, more and more eyes have been cleansed of the films which prevented them from seeing it; until now the weight and number of those who refuse to be the prey of verbal mystifications has begun to tell in practical life.

It was inevitable that a conflict should arise between Agnosticism and Theology; or rather, I ought to say, between Agnosticism and Ecclesiasticism. For Theology, the science, is one thing; and Ecclesiasticism, the championship of a foregone conclusion [1] as to the truth of a particular

---

[1] "Let us maintain, before we have proved. This seeming paradox is the secret of happiness" (Dr. Newman : Tract 85, p. 85).

form of Theology, is another. With scientific
Theology, Agnosticism has no quarrel. On the
contrary, the Agnostic, knowing too well the
influence of prejudice and idiosyncrasy, even on
those who desire most earnestly to be impartial,
can wish for nothing more urgently than that the
scientific theologian should not only be at perfect
liberty to thresh out the matter in his own
fashion ; but that he should, if he can, find flaws
in the Agnostic position ; and, even if demonstration is not to be had, that he should put, in their
full force, the grounds of the conclusions he thinks
probable. The scientific theologian admits the
Agnostic principle, however widely his results
may differ from those reached by the majority of
Agnostics.

But, as between Agnosticism and Ecclesiasticism, or, as our neighbours across the Channel
call it, Clericalism, there can be neither peace nor
truce. The Cleric asserts that it is morally wrong
not to believe certain propositions, whatever the
results of a strict scientific investigation of the
evidence of these propositions. He tells us " that
religious error is, in itself, of an immoral nature." [1]
He declares that he has prejudged certain conclusions, and looks upon those who show cause
for arrest of judgment as emissaries of Satan. It
necessarily follows that, for him, the attainment
of faith, not the ascertainment of truth, is the

[1] Dr. Newman, *Essay on Development,* p. 357.

highest aim of mental life. And, on careful
analysis of the nature of this faith, it will too
often be found to be, not the mystic process of
unity with the Divine, understood by the religious
enthusiast; but that which the candid simplicity
of a Sunday scholar once defined it to be.
"Faith," said this unconscious plagiarist of
Tertullian, "is the power of saying you believe
things which are incredible."

Now I, and many other Agnostics, believe that
faith, in this sense, is an abomination; and though
we do not indulge in the luxury of self-righteous-
ness so far as to call those who are not of our way
of thinking hard names, we do feel that the
disagreement between ourselves and those who
hold this doctrine is even more moral than
intellectual. It is desirable there should be an
end of any mistakes on this topic. If our clerical
opponents were clearly aware of the real state of
the case, there would be an end of the curious
delusion, which often appears between the lines
of their writings, that those whom they are so
fond of calling "Infidels" are people who not
only ought to be, but in their hearts are, ashamed
of themselves. It would be discourteous to do
more than hint the antipodal opposition of this
pleasant dream of theirs to facts.

The clerics and their lay allies commonly tell
us, that if we refuse to admit that there is good
ground for expressing definite convictions about

certain topics, the bonds of human society will dissolve and mankind lapse into savagery. There are several answers to this assertion. One is that the bonds of human society were formed without the aid of their theology; and, in the opinion of not a few competent judges, have been weakened rather than strengthened by a good deal of it. Greek science, Greek art, the ethics of old Israel, the social organisation of old Rome, contrived to come into being, without the help of any one who believed in a single distinctive article of the simplest of the Christian creeds. The science the art, the jurisprudence, the chief political and social theories, of the modern world have grown out of those of Greece and Rome—not by favour of, but in the teeth of, the fundamental teachings of early Christianity, to which science, art, and any serious occupation with the things of this world, were alike despicable.

Again, all that is best in the ethics of the modern world, in so far as it has not grown out of Greek thought, or Barbarian manhood, is the direct development of the ethics of old Israel. There is no code of legislation, ancient or modern, at once so just and so merciful, so tender to the weak and poor, as the Jewish law; and, if the Gospels are to be trusted, Jesus of Nazareth himself declared that he taught nothing but that which lay implicitly, or explicitly, in the religious and ethical system of his people.

And the scribe said unto him, Of a truth, Teacher, thou hast well said that he is one ; and there is none other but he and to love him with all the heart, and with all the understanding, and with all the strength, and to love his neighbour as himself, is much more than all whole burnt offerings and sacrifices. (Mark xii. 32, 33.)

Here is the briefest of summaries of the teaching of the prophets of Israel of the eighth century ; does the Teacher, whose doctrine is thus set forth in his presence, repudiate the exposition ? Nay ; we are told, on the contrary, that Jesus saw that he " answered discreetly," and replied, " Thou art not far from the kingdom of God."

So that I think that even if the creeds, from the so-called " Apostles ' " to the so-called " Athanasian," were swept into oblivion ; and even if the human race should arrive at the conclusion that, whether a bishop washes a cup or leaves it unwashed, is not a matter of the least consequence, it will get on very well. The causes which have led to the development of morality in mankind, which have guided or impelled us all the way from the savage to the civilised state, will not cease to operate because a number of ecclesiastical hypotheses turn out to be baseless. And, even if the absurd notion that morality is more the child of speculation than of practical necessity and inherited instinct, had any foundation ; if all the world is going to thieve, murder, and otherwise misconduct itself as soon as it discovers that certain portions of ancient history are mythical; what is the relevance of such arguments to any one who holds by the Agnostic principle?

Surely, the attempt to cast out Beelzebub by the aid of Beelzebub is a hopeful procedure as compared to that of preserving morality by the aid of immorality. For I suppose it is admitted that an Agnostic may be perfectly sincere, may be competent, and may have studied the question at issue with as much care as his clerical opponents. But, if the Agnostic really believes what he says, the "dreadful consequence" argufier (consistently, I admit, with his own principles) virtually asks him to abstain from telling the truth, or to say what he believes to be untrue, because of the supposed injurious consequences to morality. "Beloved brethren, that we may be spotlessly moral, before all things let us lie," is the sum total of many an exhortation addressed to the "Infidel." Now, as I have already pointed out, we cannot oblige our exhorters. We leave the practical application of the convenient doctrines of "Reserve" and "Nonnatural interpretation" to those who invented them.

I trust that I have now made amends for any ambiguity, or want of fulness, in my previous exposition of that which I hold to be the essence of the Agnostic doctrine. Henceforward, I might hope to hear no more of the assertion that we are necessarily Materialists, Idealists, Atheists,

Theists, or any other *ists*, if experience had led me
to think that the proved falsity of a statement
was any guarantee against its repetition. And
those who appreciate the nature of our position
will see, at once, that when Ecclesiasticism
declares that we ought to believe this, that, and
the other, and are very wicked if we don't, it is
impossible for us to give any answer but this:
We have not the slightest objection to believe
anything you like, if you will give us good grounds
for belief; but, if you cannot, we must respectfully
refuse, even if that refusal should wreck morality
and insure our own damnation several times over.
We are quite content to leave that to the decision
of the future. The course of the past has impressed us with the firm conviction that no good
ever comes of falsehood, and we feel warranted in
refusing even to experiment in that direction.

In the course of the present discussion it has
been asserted that the "Sermon on the Mount"
and the "Lord's Prayer" furnish a summary and
condensed view of the essentials of the teaching of
Jesus of Nazareth, set forth by himself. Now this
supposed *Summa* of Nazarene theology distinctly
affirms the existence of a spiritual world, of a
Heaven, and of a Hell of fire; it teaches the
Fatherhood of God and the malignity of the
Devil; it declares the superintending providence of
the former and our need of deliverance from the

machinations of the latter; it affirms the fact of
demoniac possession and the power of casting out
devils by the faithful. And, from these premises,
the conclusion is drawn, that those Agnostics who
deny that there is any evidence of such a character
as to justify certainty, respecting the existence and
the nature of the spiritual world, contradict the
express declarations of Jesus. I have replied to
this argumentation by showing that there is strong
reason to doubt the historical accuracy of the
attribution to Jesus of either the "Sermon on
the Mount" or the "Lord's Prayer"; and, therefore, that the conclusion in question is not
warranted, at any rate, on the grounds set
forth.

But, whether the Gospels contain trustworthy
statements about this and other alleged historical
facts or not, it is quite certain that from them,
taken together with the other books of the New
Testament, we may collect a pretty complete
exposition of that theory of the spiritual world
which was held by both Nazarenes and Christians;
and which was undoubtedly supposed by them to
be fully sanctioned by Jesus, though it is just as
clear that they did not imagine it contained any
revelation by him of something heretofore unknown. If the pneumatological doctrine which
pervades the whole New Testament is nowhere
systematically stated, it is everywhere assumed.
The writers of the Gospels and of the Acts take it

for granted, as a matter of common knowledge;
and it is easy to gather from these sources a
series of propositions, which only need arrangement to form a complete system.

In this system, Man is considered to be a
duality formed of a spiritual element, the soul;
and a corporeal[1] element, the body. And this
duality is repeated in the Universe, which consists
of a corporeal world embraced and interpenetrated
by a spiritual world. The former consists of the
earth, as its principal and central constituent, with
the subsidiary sun, planets, and stars. Above the
earth is the air, and below is the watery abyss.
Whether the heaven, which is conceived to be
above the air, and the hell in, or below, the subterranean deeps, are to be taken as corporeal or
incorporeal is not clear. However this may be,
the heaven and the air, the earth and the abyss,
are peopled by innumerable beings analogous in
nature to the spiritual element in man, and these
spirits are of two kinds, good and bad. The chief
of the good spirits, infinitely superior to all the
others, and their creator, as well as the creator of
the corporeal world and of the bad spirits, is God.

---

[1] It is by no means to be assumed that "spiritual" and "corporeal" are exact equivalents of "immaterial" and "material"
in the minds of ancient speculators on these topics. The
"spiritual body" of the risen dead (1 Cor. xv.) is not the
"natural" "flesh and blood" body. Paul does not teach the
resurrection of the body in the ordinary sense of the word
"body"; a fact, often overlooked, but pregnant with many
consequences.

His residence is heaven, where he is surrounded by the ordered hosts of good spirits ; his angels, or messengers, and the executors of his will throughout the universe.

On the other hand, the chief of the bad spirits is Satan, *the* devil *par excellence.* He and his company of demons are free to roam through all parts of the universe, except the heaven. These bad spirits are far superior to man in power and subtlety ; and their whole energies are devoted to bringing physical and moral evils upon him, and to thwarting, so far as their power goes, the benevolent intentions of the Supreme Being. In fact, the souls and bodies of men form both the theatre and the prize of an incessant warfare between the good and the evil spirits—the powers of light and the powers of darkness. By leading Eve astray, Satan brought sin and death upon mankind. As the gods of the heathen, the demons are the founders and maintainers of idolatry ; as the " powers of the air " they afflict mankind with pestilence and famine ; as " unclean spirits " they cause disease of mind and body.

The significance of the appearance of Jesus, in the capacity of the Messiah, or Christ, is the reversal of the satanic work by putting an end to both sin and death. He announces that the kingdom of God is at hand, when the " Prince of this world " shall be finally " cast out " (John xii. 31) from the cosmos, as Jesus, during his earthly

career, cast him out from individuals. Then will
Satan and all his devilry, along with the wicked
whom they have seduced to their destruction, be
hurled into the abyss of unquenchable fire—there
to endure continual torture, without a hope of
winning pardon from the merciful God, their
Father; or of moving the glorified Messiah to one
more act of pitiful intercession; or even of
interrupting, by a momentary sympathy with
their wretchedness, the harmonious psalmody of
their brother angels and men, eternally lapped in
bliss unspeakable.

The straitest Protestant, who refuses to admit
the existence of any source of Divine truth,
except the Bible, will not deny that every point
of the pneumatological theory here set forth has
ample scriptural warranty. The Gospels, the
Acts, the Epistles, and the Apocalypse assert the
existence of the devil, of his demons and of Hell,
as plainly as they do that of God and his angels
and Heaven. It is plain that the Messianic and
the Satanic conceptions of the writers of these
books are the obverse and the reverse of the same
intellectual coinage. If we turn from Scripture
to the traditions of the Fathers and the confessions of the Churches, it will appear that, in this
one particular, at any rate, time has brought
about no important deviation from primitive
belief. From Justin onwards, it may often be a
fair question whether God, or the devil, occupies

a larger share of the attention of the Fathers. It is the devil who instigates the Roman authorities to persecute; the gods and goddesses of paganism are devils, and idolatry itself is an invention of Satan; if a saint falls away from grace, it is by the seduction of the demon; if heresy arises, the devil has suggested it; and some of the Fathers[1] go so far as to challenge the pagans to a sort of exorcising match, by way of testing the truth of Christianity. Mediæval Christianity is at one with patristic, on this head. The masses, the clergy, the theologians, and the philosophers alike, live and move and have their being in a world full of demons, in which sorcery and possession are everyday occurrences. Nor did the Reformation make any difference. Whatever else Luther assailed, he left the traditional demonology untouched; nor could any one have entertained a more hearty and uncompromising belief in the devil, than he and, at a later period, the Calvinistic fanatics of New England did. Finally, in these last years of the nineteenth century, the demonological hypotheses of the first century are, explicitly or implicitly, held and occasionally acted upon by the immense majority of Christians of all confessions.

---

[1] Tertullian (*Apolog. adv. Gentes*, cap. xxiii.) thus challenges the Roman authorities : let them bring a possessed person into the presence of a Christian before their tribunal ; and if the demon does not confess himself to be such, on the order of the Christian, let the Christian be executed out of hand.

Only here and there has the progress of scientific thought, outside the ecclesiastical world, so far affected Christians, that they and their teachers fight shy of the demonology of their creed. They are fain to conceal their real disbelief in one half of Christian doctrine by judicious silence about it; or by flight to those refuges for the logically destitute, accommodation or allegory. But the faithful who fly to allegory in order to escape absurdity resemble nothing so much as the sheep in the fable who—to save their lives—jumped into the pit. The allegory pit is too commodious, is ready to swallow up so much more than one wants to put into it. If the story of the temptation is an allegory; if the early recognition of Jesus as the Son of God by the demons is an allegory; if the plain declaration of the writer of the first Epistle of John (iii. 8), " To this end was the Son of God manifested, that He might destroy the works of the devil," is allegorical, then the Pauline version of the Fall may be allegorical, and still more the words of consecration of the Eucharist, or the promise of the second coming; in fact, there is not a dogma of ecclesiastical Christianity the scriptural basis of which may not be whittled away by a similar process.

As to accommodation, let any honest man who can read the New Testament ask himself whether Jesus and his immediate friends and disciples can be dishonoured more grossly than by the supposition that they said and did that which is attributed to them; while, in reality, they disbelieved in Satan and his demons, in possession and in exorcism ?[1]

An eminent theologian has justly observed that we have no right to look at the propositions of the Christian faith with one eye open and the other shut. (Tract 85, p. 29.) It really is not permissible to see, with one eye, that Jesus is affirmed to declare the personality and the Fatherhood of God, His loving providence and His accessibility to prayer; and to shut the other to the no less definite teaching ascribed to Jesus, in regard to the personality and the misanthropy of the devil, his malignant watchfulness, and his subjection to exorcistic formulæ and rites.   Jesus is made to say that the devil " was a murderer from the beginning " (John viii. 44) by the same authority as that upon which we depend for his asserted declaration that " God is a spirit " (John iv. 24).

To those who admit the authority of the famous Vincentian dictum that the doctrine which has been held " always, everywhere, and by all " is to be received as authoritative, the demonology must possess a higher sanction than any other Christian dogma, except, perhaps, those of the Resurrection and of the Messiahship of Jesus;

---

[1] See the expression of orthodox opinion upon the "accommodation" subterfuge already cited above, p. 217.

for it would be difficult to name any other points
of doctrine on which the Nazarene does not differ
from the Christian, and the different historical
stages and contemporary subdivisions of Christianity from one another. And, if the demonology is accepted, there can be no reason for
rejecting all those miracles in which demons play
a part. The Gadarene story fits into the general
scheme of Christianity; and the evidence for
" Legion " and their doings is just as good as any
other in the New Testament for the doctrine
which the story illustrates.

It was with the purpose of bringing this great
fact into prominence; of getting people to open
both their eyes when they look at Ecclesiasticism;
that I devoted so much space to that miraculous
story which happens to be one of the best types
of its class. And I could not wish for a better
justification of the course I have adopted, than
the fact that my heroically consistent adversary
has declared his implicit belief in the Gadarene
story and (by necessary consequence) in the
Christian demonology as a whole. It must be
obvious, by this time, that, if the account of the
spiritual world given in the New Testament, professedly on the authority of Jesus, is true, then
the demonological half of that account must be
just as true as the other half. And, therefore,
those who question the demonology, or try to
explain it away, deny the truth of what Jesus

said, and are, in ecclesiastical terminology, "Infidels" just as much as those who deny the spirituality of God. This is as plain as anything can well be, and the dilemma for my opponent was either to assert that the Gadarene pig-bedevilment actually occurred, or to write himself down an "Infidel." As was to be expected, he chose the former alternative; and I may express my great satisfaction at finding that there is one spot of common ground on which both he and I stand. So far as I can judge, we are agreed to state one of the broad issues between the consequences of agnostic principles (as I draw them), and the consequences of ecclesiastical dogmatism (as he accepts it), as follows.

Ecclesiasticism says: The demonology of the Gospels is an essential part of that account of that spiritual world, the truth of which it declares to be certified by Jesus.

Agnosticism (*me judice*) says: There is no good evidence of the existence of a demoniac spiritual world, and much reason for doubting it.

Hereupon the ecclesiastic may observe: Your doubt means that you disbelieve Jesus; therefore you are an "Infidel" instead of an "Agnostic." To which the agnostic may reply: No; for two reasons: first, because your evidence that Jesus said what you say he said is worth very little; and secondly, because a man may be an agnostic, in the sense of admitting he has no positive knowledge, and yet consider that he has more or less probable ground for accepting any given hypothesis about the spiritual world. Just as a man may frankly declare that he has no means of knowing whether the planets generally are inhabited or not, and yet may think one of the two possible hypotheses more likely than the other, so he may admit that he has no means of knowing anything about the spiritual world, and yet may think one or other of the current views on the subject, to some extent, probable.

The second answer is so obviously valid that it needs no discussion. I draw attention to it simply in justice to those agnostics who may attach greater value than I do to any sort of pneumatological speculations; and not because I wish to escape the responsibility of declaring that, whether Jesus sanctioned the demonological part of Christianity or not, I unhesitatingly reject it. The first answer, on the other hand, opens up the whole question of the claim of the biblical and other sources, from which hypotheses concerning the spiritual world are derived, to be regarded as unimpeachable historical evidence as to matters of fact.

Now, in respect of the trustworthiness of the Gospel narratives, I was anxious to get rid of the common assumption that the determination of the authorship and of the dates of these works is a matter of fundamental importance. That assump-

tion is based upon the notion that what contemporary witnesses say must be true, or, at least, has
always a *primâ facie* claim to be so regarded ; so
that if the writers of any of the Gospels were
contemporaries of the events (and still more if
they were in the position of eye-witnesses) the
miracles they narrate must be historically true,
and, consequently, the demonology which they
involve must be accepted. But the story of the
" Translation of the blessed martyrs Marcellinus
and Petrus," and the other considerations (to
which endless additions might have been made
from the Fathers and the mediæval writers) set
forth in a preceding essay, yield, in my judgment,
satisfactory proof that, where the miraculous is
concerned, neither considerable intellectual ability,
nor undoubted honesty, nor knowledge of the
world, nor proved faithfulness as civil historians,
nor profound piety, on the part of eye witnesses
and contemporaries, affords any guarantee of the
objective truth of their statements, when we know
that a firm belief in the miraculous was ingrained
in their minds, and was the pre-supposition of
their observations and reasonings.

Therefore, although it be, as I believe, demonstrable that we have no real knowledge of the
authorship, or of the date of composition of the
Gospels, as they have come down to us, and
that nothing better than more or less probable
guesses can be arrived at on that subject, I have

not cared to expend any space on the question. It will be admitted, I suppose, that the authors of the works attributed to Matthew, Mark, Luke, and John, whoever they may be, are personages whose capacity and judgment in the narration of ordinary events are not quite so well certified as those of Eginhard ; and we have seen what the value of Eginhard's evidence is when the miraculous is in question.

I have been careful to explain that the arguments which I have used in the course of this discussion are not new ; that they are historical and have nothing to do with what is commonly called science ; and that they are all, to the best of my belief, to be found in the works of theologians of repute.

The position which I have taken up, that the evidence in favour of such miracles as those recorded by Eginhard, and consequently of mediæval demonology, is quite as good as that in favour of such miracles as the Gadarene, and consequently of Nazarene demonology, is none of my discovery. Its strength was, wittingly or unwittingly, suggested, a century and a half ago, by a theological scholar of eminence ; and it has been, if not exactly occupied, yet so fortified with bastions and redoubts by a living ecclesiastical Vauban, that, in my judgment, it has been rendered impregnable. In the early part of the last

century, the ecclesiastical mind in this country
was much exercised by the question, not exactly
of miracles, the occurrence of which in biblical
times was axiomatic, but by the problem: When
did miracles cease? Anglican divines were quite
sure that no miracles had happened in their day,
nor for some time past; they were equally sure
that they happened sixteen or seventeen centuries
earlier. And it was a vital question for them to
determine at what point of time, between this
terminus a quo and that *terminus ad quem*,
miracles came to an end.

The Anglicans and the Romanists agreed in
the assumption that the possession of the gift of
miracle-working was *primâ facie* evidence of the
soundness of the faith of the miracle-workers.
The supposition that miraculous powers might be
wielded by heretics (though it might be supported
by high authority) led to consequences too frightful to be entertained by people who were busied
in building their dogmatic house on the sands of
early Church history. If, as the Romanists maintained, an unbroken series of genuine miracles
adorned the records of their Church, throughout
the whole of its existence, no Anglican could
lightly venture to accuse them of doctrinal cor-
ruption. Hence, the Anglicans, who indulged in
such accusations, were bound to prove the modern,
the mediæval Roman, and the later Patristic,
miracles false; and to shut off the wonder-working

power from the Church at the exact point of
time when Anglican doctrine ceased and Roman
doctrine began.   With a little adjustment—a
squeeze here and a pull there—the Christianity
of the first three or four centuries might be made
to fit, or seem to fit, pretty well into the Anglican
scheme.   So the miracles, from Justin say to
Jerome, might be recognised; while, in later
times, the Church having become "corrupt"—
that is to say, having pursued one and the same
line of development further than was pleasing to
Anglicans—its alleged miracles must needs be
shams and impostures.

Under these circumstances, it may be imagined
that the establishment of a scientific frontier
between the earlier realm of supposed fact and
the later of asserted delusion, had its difficulties;
and torrents of theological special pleading about
the subject flowed from clerical pens; until that
learned and acute Anglican divine, Conyers
Middleton, in his "Free Inquiry," tore the sophistical web they had laboriously woven to pieces, and
demonstrated that the miracles of the patristic
age, early and late, must stand or fall together,
inasmuch as the evidence for the later is just as
good as the evidence for the earlier wonders.   If
the one set are certified by contemporaneous
witnesses of high repute, so are the other; and,
in point of probability, there is not a pin to choose
between the two.   That is the solid and irrefrag-

able result of Middleton's contribution to the subject. But the Free Inquirer's freedom had its limits; and he draws a sharp line of demarcation between the patristic and the New Testament miracles—on the professed ground that the accounts of the latter, being inspired, are out of the reach of criticism.

A century later, the question was taken up by another divine, Middleton's equal in learning and acuteness, and far his superior in subtlety and dialetic skill; who, though an Anglican, scorned the name of Protestant; and, while yet a Churchman, made it his business, to parade, with infinite skill, the utter hollowness of the arguments of those of his brother Churchmen who dreamed that they could be both Anglicans and Protestants. The argument of the " Essay on the Miracles recorded in the Ecclesiastical History of the Early Ages"[1] by the present [1889] Roman Cardinal, but then Anglican Doctor, John Henry Newman, is compendiously stated by himself in the following passage :—

> If the miracles of Church history cannot be defended by the arguments of Leslie, Lyttleton, Paley, or Douglas, how many of the Scripture miracles satisfy their conditions ? (p. cvii).

And, although the answer is not given in so many words, little doubt is left on the mind of the

---

[1] I quote the first edition (1843). A second edition appeared in 1870. Tract 85 of the *Tracts for the Times* should be read with this *Essay*. If I were called upon to compile a Primer of "Infidelity," I think I should save myself trouble by making a selection from these works, and from the *Essay on Development* by the same author.

reader, that, in the mind of the writer, it is : None.
In fact, this conclusion is one which cannot be
resisted, if the argument in favour of the Scripture
miracles is based upon that which laymen,
whether lawyers, or men of science, or historians,
or ordinary men of affairs, call evidence. But
there is something really impressive in the
magnificent contempt. with which, at times, Dr.
Newman sweeps aside alike those who offer and
those who demand such evidence.

> Some infidel authors advise us to accept no miracles which
> would not have a verdict in their favour in a court of justice ;
> that is, they employ against Scripture a weapon which Pro-
> testants would confine to attacks upon the Church ; as if moral
> and religious questions required legal proof, and evidence were
> the test of truth [1] (p. cvii).

" As if evidence were the test of truth " !—although
the truth in question is the occurrence, or the
non-occurrence, of certain phenomena at a certain
time and in a certain place. This sudden revelation
of the great gulf fixed between the ecclesiastical
and the scientific mind is enough to take away
the breath of any one unfamiliar with the clerical
organon. As if, one may retort, the assumption
that miracles may, or have, served a moral or a
religious end, in any way alters the fact that they
profess to be historical events, things that actually

---

[1] Yet, when it suits his purpose, as in the Introduction to the
Essay on Development, Dr. Newman can demand strict evidence
in religious questions as sharply as any "infidel author" ; and
he can even profess to yield to its force (*Essay on Miracles*, 1870 ;
note, p. 391).

happened; and, as such, must needs be exactly
those subjects about which evidence is appropriate
and legal proofs (which are such merely because
they afford adequate evidence) may be justly
demanded. The Gadarene miracle either happened, or it did not. Whether the Gadarene
"question" is moral or religious, or not, has
nothing to do with the fact that it is a purely
historical question whether the demons said what
they are declared to have said, and the devil-
possessed pigs did, or did not, rush over the heights
bounding the Lake of Gennesaret on a certain day
of a certain year, after A.D. 26 and before A.D. 36 :
for vague and uncertain as New Testament
chronology is, I suppose it may be assumed that
the event in question, if it happened at all, took
place during the procuratorship of Pilate. If that
is not a matter about which evidence ought to be
required, and not only legal, but strict scientific
proof demanded by sane men who are asked to
believe the story—what is ? Is a reasonable
being to be seriously asked to credit statements,
which, to put the case gently, are not exactly
probable, and on the acceptance or rejection of
which his whole view of life may depend, without
asking for as much "legal" proof as would send
an alleged pickpocket to gaol, or as would suffice
to prove the validity of a disputed will ?

"Infidel authors" (if, as I am assured, I may
answer for them) will decline to waste time on

mere darkenings of counsel of this sort; but to those Anglicans who accept his premises, Dr. Newman is a truly formidable antagonist. What, indeed, are they to reply when he puts the very pertinent question :—

whether persons who not merely question, but prejudge the Ecclesiastical miracles on the ground of their want of resemblance, whatever that be, to those contained in Scripture—as if the Almighty could not do in the Christian Church what He had not already done at the time of its foundation, or under the Mosaic Covenant—whether such reasoners are not siding with the sceptic,

and

whether it is not a happy inconsistency by which they continue to believe the Scriptures while they reject the Church [1] (p. liii).

Again, I invite Anglican orthodoxy to consider this passage :—

the narrative of the combats of St. Antony with evil spirits, is a development rather than a contradiction of revelation, viz. of such texts as speak of Satan being cast out by prayer and fasting. To be shocked, then, at the miracles of Ecclesiastical history, or to ridicule them for their strangeness, is no part of a scriptural philosophy (pp. liii-liv).

Further on, Dr. Newman declares that it has been admitted

that a distinct line can be drawn in point of character and cir cumstance between the miracles of Scripture and of Church

---

[1] Compare Tract 85, p. 110 ; " I am persuaded that were men but consistent who oppose the Church doctrines as being unscriptural, they would vindicate the Jews for rejecting the Gospel."

history; but this is by no means the case (p. lv). . . . specimens are not wanting in the history of the Church, of miracles
as awful in their character and as momentous in their effects as
those which are recorded in Scripture. The fire interrupting
the rebuilding of the Jewish temple, and the death of Arius, are
instances, in Ecclesiastical history, of such solemn events. On
the other hand, difficult instances in the Scripture history are
such as these : the serpent in Eden, the Ark, Jacob's vision for
the multiplication of his cattle, the speaking of Balaam's ass,
the axe swimming at Elisha's word, the miracle on the swine,
and various instances of prayers or prophecies, in which, as in
that of Noah's blessing and curse, words which seem the result
of private feeling are expressly or virtually ascribed to a Divine
suggestion (p. lvi).

Who is to gainsay our ecclesiastical authority
here ? "Infidel authors" might be accused of a
wish to ridicule the Scripture miracles by putting
them on a level with the remarkable story about
the fire which stopped the rebuilding of the
Temple, or that about the death of Arius—but
Dr. Newman is above suspicion. The pity is that
his list of what he delicately terms "difficult"
instances is so short. Why omit the manufacture
of Eve out of Adam's rib, on the strict historical
accuracy of which the chief argument of the
defenders of an iniquitous portion of our present
marriage law depends ? Why leave out the
account of the "Bene Elohim" and their gallan-
tries, on which a large part of the worst practices
of the mediæval inquisitors into witchcraft was
based ? Why forget the angel who wrestled with
Jacob, and, as the account suggests, somewhat

over-stepped the bound of fair play, at the end of
the struggle? Surely, we must agree with Dr.
Newman that, if all these camels have gone down,
it savours of affectation to strain at such gnats as
the sudden ailment of Arius in the midst of his
deadly, if prayerful,[1] enemies ; and the fiery explo-
sion which stopped the Julian building operations.
Though the *words* of the "Conclusion" of the
"Essay on Miracles" may, perhaps, be quoted
against me, I may express my satisfaction at finding
myself in substantial accordance with a theologian
above all suspicion of heterodoxy. With all my
heart, I can declare my belief that there is just as
good reason for believing in the miraculous slaying of the man who fell short of the Athanasian
power of affirming contradictories, with respect to
the nature of the Godhead, as there is for believing
in the stories of the serpent and the ark told in
Genesis, the speaking of Balaam's ass in Numbers,
or the floating of the axe, at Elisha's order, in the
second book of Kings.

It is one of the peculiarities of a really sound

---

[1] According to Dr. Newman, "This prayer [that of Bishop
Alexander, who begged God to 'take Arius away'] is said to
have been offered about 3 P.M. on the Saturday ; that same
evening Arius was in the great square of Constantine, when he
was suddenly seized with indisposition" (p. clxx). The
"infidel" Gibbon seems to have dared to suggest that "an
option between poison and miracle" is presented by this
case ; and, it must be admitted, that, if the Bishop had been
within the reach of a modern police magistrate, things might
have gone hardly with him. Modern "Infidels," possessed of a

argument that it is susceptible of the fullest
development; and that it sometimes leads to conclusions unexpected by those who employ it. To
my mind, it is impossible to refuse to follow Dr.
Newman when he extends his reasoning, from the
miracles of the patristic and mediæval ages back-
ward in time, as far as miracles are recorded.
But, if the rules of logic are valid, I feel compelled to extend the argument forwards to the
alleged Roman miracles of the present day, which
Dr. Newman might not have admitted, but which
Cardinal Newman may hardly reject. Beyond
question, there is as good, or perhaps better,
evidence for the miracles worked by our Lady of
Lourdes, as there is for the floating of Elisha's axe,
or the speaking of Balaam's ass. But we must go
still further; there is a modern system of thauma-
turgy and demonology which is just as well
certified as the ancient.[1] Veracious, excellent,

slight knowledge of chemistry, are not unlikely, with no less
audacity, to suggest an "option between fire-damp and miracle"
in seeking for the cause of the fiery outburst at Jerusalem.

[1] A writer in a spiritualist journal takes me roundly to task
for venturing to doubt the historical and literal truth of the
Gadarene story. The following passage in his letter is worth
quotation : " Now to the materialistic and scientific mind, to the
uninitiated in spiritual verities, certainly this story of the
Gadarene or Gergesene swine presents insurmountable difficulties ;
it seems grotesque and nonsensical. To the experienced, trained,
and cultivated Spiritualist this miracle is, as I am prepared to
show, one of the most instructive, the most profoundly useful,
and the most beneficent which Jesus ever wrought in
the whole course of His pilgrimage of redemption on earth."
Just so. And the first page of this same journal presents the
following advertisement, among others of the same kidney :—
   "To WEALTHY SPIRITUALISTS.—A Lady Medium of tried

sometimes learned and acute persons, even philosophers of no mean pretensions, testify to the
"levitation" of bodies much heavier than Elisha's
axe; to the existence of "spirits" who, to the
mere tactile sense, have been indistinguishable
from flesh and blood; and, occasionally, have
wrestled with all the vigour of Jacob's opponent;
yet, further, to the speech, in the language of raps,
of spiritual beings, whose discourses, in point of
coherence and value, are far inferior to that of
Balaam's humble but sagacious steed.   I have not
the smallest doubt that, if these were persecuting
times, there is many a worthy "spiritualist" who
would cheerfully go to the stake in support of his
pneumatological faith; and furnish evidence, after
Paley's own heart, in proof of the truth of his
doctrines.   Not a few modern divines, doubtless
struck by the impossibility of refusing the spiritualist evidence, if the ecclesiastical evidence is
accepted, and deprived of any *à priori* objection
by their implicit belief in Christian Demonology,
show themselves ready to take poor Sludge
seriously, and to believe that he is possessed by
other devils than those of need, greed, and vainglory.

Under these circumstances, it was to be

power wishes to meet with an elderly gentleman who would be
willing to give her a comfortable home and maintenance in
Exchange for her Spiritualistic services, as her guides consider
her health is too delicate for public sittings : London preferred.—
Address 'Mary,' Office of *Light*."

Are we going back to the days of the Judges, when wealthy
Micah set up his private ephod, teraphim, and Levite ?

expected, though it is none the less interesting to note the fact, that the arguments of the latest school of "spiritualists" present a wonderful family likeness to those which adorn the subtle disquisitions of the advocate of ecclesiastical miracles of forty years ago. It is unfortunate for the "spiritualists" that, over and over again, cele-brated and trusted media, who really, in some respects, call to mind the Montanist[1] and gnostic seers of the second century, are either proved in courts of law to be fraudulent impostors; or, in sheer weariness, as it would seem, of the honest dupes who swear by them, spontaneously confess their long-continued iniquities, as the Fox women did the other day in New York.[2]   But, whenever a catastrophe of this kind takes place, the believers are no wise dismayed by it. They freely admit that not only the media, but the spirits whom they summon, are sadly apt to lose sight of the elemen-tary principles of right and wrong; and they triumphantly ask : How does the occurrence of

[1] Consider Tertullian's "sister" ("hodie apud nos"), who conversed with angels, saw and heard mysteries, knew men's thoughts, and prescribed medicine for their bodies (*De Anima*, cap. 9). Tertullian tells us that this woman saw the soul as corporeal, and described its colour and shape. The "infidel" will probably be unable to refrain from insulting the memory of the ecstatic saint by the remark, that Tertullian's known views about the corporeality of the soul may have had some-thing to do with the remarkable perceptive powers of the Montanist medium, in whose revelations of the spiritual world he took such profound interest.

[2] See the New York *World* for Sunday, 21st October, 1888 ; and the *Report of the Seybert Commission*, Philadelphia, 1887.

drawing a line in the series that might be set out of plausibly attested cases of spiritual intervention. If one is true, all may be true; if one is false, all may be false.

This is, to my mind, the inevitable result of that method of reasoning which is applied to the confutation of Protestantism, with so much success, by one of the acutest and subtlest disputants who have ever championed Ecclesiasticism —and one cannot put his claims to acuteness and subtlety higher.

> . . . the Christianity of history is not Protestantism. If ever there were a safe truth it is this. . . . "To be deep in history is to cease to be a Protestant."[1]

I have not a shadow of doubt that these anti-Protestant epigrams are profoundly true. But I have as little that, in the same sense, the "Christianity of history is not" Romanism; and that to be deeper in history is to cease to be a Romanist. The reasons which compel my doubts about the compatibility of the Roman doctrine, or any other form of Catholicism, with history, arise out of exactly the same line of argument as that adopted by Dr. Newman in the famous essay which I have just cited. If, with one hand, Dr. Newman has destroyed Protestantism, he has

---

[1] *An Essay on the Development of Christian Doctrine*, by J. H. Newman, D.D., pp. 7 and 8. (1878.)

annihilated Romanism with the other; and the total result of his ambidextral efforts is to shake Christianity to its foundations. Nor was any one better aware that this must be the inevitable result of his arguments—if the world should refuse to accept Roman doctrines and Roman miracles—than the writer of Tract 85.

Dr. Newman made his choice and passed over to the Roman Church half a century ago. Some of those who were essentially in harmony with his views preceded, and many followed him. But many remained; and, as the quondam Puseyite and present Ritualistic party, they are continuing that work of sapping and mining the Protestantism of the Anglican Church which he and his friends so ably commenced. At the present time, they have no little claim to be considered victorious all along the line. I am old enough to recollect the small beginnings of the Tractarian party; and I am amazed when I consider the present position of their heirs. Their little leaven has leavened, if not the whole, yet a very large lump of the Anglican Church; which is now pretty much of a preparatory school for Papistry. So that it really behoves Englishmen (who, as I have been informed by high authority, are all legally, members of the State Church, if they profess to belong to no other sect) to wake up to what that powerful organisation is about, and whither it is tending. On this point, the writings

of Dr. Newman, while he still remained within
the Anglican fold, are a vast store of the best
and the most authoritative information.   His
doctrines on Ecclesiastical miracles and on
Development are the corner-stones of the Tractarian fabric.   He believed that his arguments led
either Romeward, or to what ecclesiastics call
" Infidelity," and I call Agnosticism.   I believe
that he was quite right in this conviction ; but
while he chooses the one alternative, I choose the
other ; as he rejects Protestantism on the ground
of its incompatibility with history, so, *à fortiori*,
I conceive that Romanism ought to be rejected ;
and that an impartial consideration of the evidence must refuse the authority of Jesus to
anything more than the Nazarenism of James
and Peter and John.   And let it not be supposed
that this is a mere "infidel" perversion of the facts.
No one has more openly and clearly admitted the
possibility that they may be fairly interpreted in
this way than Dr. Newman.   If, he says, there
are texts which seem to show that Jesus contemplated the evangelisation of the heathen :

. . . Did not the Apostles hear our Lord ? and what was *their*
impression from what they heard ? Is it not certain that the
Apostles did not gather this truth from His teaching ? (Tract
85, p. 63).

He said, " Preach the Gospel to every creature." These words
need have only meant " Bring all men to Christianity through
Judaism." Make them Jews, that they may enjoy Christ's
privileges, which are lodged in Judaism ; teach them those

rites and ceremonies, circumcision and the like, which hitherto have been dead ordinances, and now are living: and so the Apostles seem to have understood them (*ibid.* p. 65).

So far as Nazarenism differentiated itself from contemporary orthodox Judaism, it seems to have tended towards a revival of the ethical and religious spirit of the prophetic age, accompanied by the belief in Jesus as the Messiah, and by various accretions which had grown round Judaism subsequently to the exile. To these belong the doctrines of the Resurrection, of the Last Judgment, of Heaven and Hell; of the hierarchy of good angels; of Satan and the hierarchy of evil spirits. And there is very strong ground for believing that all these doctrines, at least in the shapes in which they were held by the post-exilic Jews, were derived from Persian and Babylonian [1] sources, and are essentially of heathen origin.

How far Jesus positively sanctioned all these indrainings of circumjacent Paganism into Judaism; how far any one has a right to declare, that the refusal to accept one or other of these doctrines, as ascertained verities, comes to the same thing as contradicting Jesus, it appears to

[1] Dr. Newman faces this question with his customary ability. "Now, I own, I am not at all solicitous to deny that this doctrine of an apostate Angel and his hosts was gained from Babylon: it might still be Divine nevertheless. God who made the prophet's ass speak, and thereby instructed the prophet, might instruct His Church by means of heathen Babylon" (Tract 85, p. 83). There seems to be no end to the apologetic burden that Balaam's ass can carry.

me not easy to say. But it is hardly less difficult
to conceive that he could have distinctly nega-
tived any of them; and, more especially, that
demonology which has been accepted by the
Christian Churches, in every age and under all
their mutual antagonisms. But, I repeat my
conviction that, whether Jesus sanctioned the
demonology of his time and nation or not, it is
doomed. The future of Christianity, as a dogmatic system and apart from the old Israelitish
ethics which it has appropriated and developed,
lies in the answer which mankind will eventually
give to the question, whether they are prepared to
believe such stories as the Gadarene and the
pneumatological hypotheses which go with it, or
not. My belief is they will decline to do anything of the sort, whenever and wherever their
minds have been disciplined by science. And
that discipline must, and will, at once follow and
lead the footsteps of advancing civilisation.

The preceding pages were written before I
became acquainted with the contents of the May
number of the "Nineteenth Century," wherein I
discover many things which are decidedly not to
my advantage. It would appear that "evasion"
is my chief resource, "incapacity for strict argument" and "rottenness of ratiocination" my main
mental characteristics, and that it is "barely
credible" that a statement which I profess to

make of my own knowledge is true. All which
things I notice, merely to illustrate the great
truth, forced on me by long experience, that it is
only from those who enjoy the blessing of a firm
hold of the Christian faith that such manifestations of meekness, patience, and charity are to be
expected.

I had imagined that no one who had read
my preceding papers, could entertain a doubt as
to my position in respect of the main issue, as
it has been stated and restated by my opponent:

an Agnosticism which knows nothing of the relation of man to
God must not only refuse belief to our Lord's most undoubted
teaching, but must deny the reality of the spiritual convictions
in which He lived.[1]

That is said to be " the simple question which is
at issue between us," and the three testimonies to
that teaching and those convictions selected are
the Sermon on the Mount, the Lord's Prayer, and
the Story of the Passion.

My answer, reduced to its briefest form, has
been: In the first place, the evidence is such that
the exact nature of the teachings and the convictions of Jesus is extremely uncertain; so that
what ecclesiastics are pleased to call a denial of
them may be nothing of the kind. And, in the
second place, if Jesus taught the demonological
system involved in the Gadarene story—if a belief

[1] *Nineteenth Century*, May 1889 (p. 701).

in that system formed a part of the spiritual convictions in which he lived and died—then I, for
my part, unhesitatingly refuse belief in that
teaching, and deny the reality of those spiritual
convictions.   And I go further and add, that,
exactly in so far as it can be proved that Jesus
sanctioned the essentially pagan demonological
theories current among the Jews of his age,
exactly in so far, for me, will his authority in
any matter touching the spiritual world be weakened.

With respect to the first half of my answer, I
have pointed out that the Sermon on the Mount,
as given in the first Gospel, is, in the opinion of
the best critics, a "mosaic work" of materials
derived from different sources, and I do not understand that this statement is challenged.   The only
other Gospel—the third—which contains something like it, makes, not only the discourse, but the
circumstances under which it was delivered, very
different.   Now, it is one thing to say that there
was something real at the bottom of the two
discourses—which is quite possible; and another
to affirm that we have any right to say what that
something was, or to fix upon any particular
phrase and declare it to be a genuine utterance.
Those who pursue theology as a science, and bring
to the study an adequate knowledge of the ways of
ancient historians, will find no difficulty in providing illustrations of my meaning.   I may supply

one which has come within range of my own
limited vision.

In Josephus's "History of the Wars of the Jews"
(chap. xix.), that writer reports a speech which
he says Herod made at the opening of a war with
the Arabians. It is in the first person, and would
naturally be supposed by the reader to be intended
for a true version of what Herod said. In the
"Antiquities," written some seventeen years later,
the same writer gives another report, also in the
first person, of Herod's speech on the same
occasion. This second oration is twice as long as
the first and, though the general tenor of the two
speeches is pretty much the same, there is hardly
any verbal identity, and a good deal of matter is
introduced into the one, which is absent from the
other. Josephus prides himself on his accuracy;
people whose fathers might have heard Herod's
oration were his contemporaries; and yet his
historical sense is so curiously undeveloped that
he can, quite innocently, perpetrate an obvious
literary fabrication; for one of the two accounts
must be incorrect. Now, if I am asked whether I
believe that Herod made some particular statement on this occasion; whether, for example, he
uttered the pious aphorism, " Where God is, there
is both multitude and courage," which is given in
the "Antiquities," but not in the "Wars," I am
compelled to say I do not know. One of the two
reports must be erroneous, possibly both are: at

any rate, I cannot tell how much of either is true. And, if some fervent admirer of the Idumean should build up a theory of Herod's piety upon Josephus's evidence that he propounded the aphorism, is it a " mere evasion " to say, in reply, that the evidence that he did utter it is worthless ?

It appears again that, adopting the tactics of Conachar when brought face to face with Hal o' the Wynd, I have been trying to get my simple-minded adversary to follow me on a wild-goose chase through the early history of Christianity, in the hope of escaping impending defeat on the main issue. But I may be permitted to point out that there is an alternative hypothesis which equally fits the facts; and that, after all, there may have been method in the madness of my supposed panic.

For suppose it to be established that Gentile Christianity was a totally different thing from the Nazarenism of Jesus and his immediate disciples; suppose it to be demonstrable that, as early as the sixth decade of our era at least, there were violent divergencies of opinion among the followers of Jesus; suppose it to be hardly doubtful that the Gospels and the Acts took their present shapes under the influence of those divergencies; suppose that their authors, and those through whose hands they passed, had notions of historical veracity not more eccentric than those which Josephus

occasionally displays: surely the chances that the
Gospels are altogether trustworthy records of the
teachings of Jesus become very slender. And,
since the whole of the case of the other side is
based on the supposition that they are accurate
records (especially of speeches, about which ancient
historians are so curiously loose), I really do venture to submit that this part of my argument bears
very seriously on the main issue; and, as ratiocination, is sound to the core.

Again, when I passed by the topic of the
speeches of Jesus on the Cross, it appears that I
could have had no other motive than the dictates
of my native evasiveness. An ecclesiastical dignitary may have respectable reasons for declining
a fencing match "in sight of Gethsemane and
Calvary"; but an ecclesiastical "Infidel"! Never.
It is obviously impossible that, in the belief that
"the greater includes the less," I, having declared
the Gospel evidence in general, as to the sayings of
Jesus, to be of questionable value, thought it needless to select for illustration of my views, those
particular instances which were likely to be most
offensive to persons of another way of thinking.
But any supposition that may have been entertained that the old familiar tones of the ecclesiastical war-drum will tempt me to engage in such
needless discussion had better be renounced. I
shall do nothing of the kind. Let it suffice that
I ask my readers to turn to the twenty-third

chapter of Luke (revised version), verse thirty-four, and he will find in the margin

Some ancient authorities omit : And Jesus said "Father, for-give them, for they know not what they do."

So that, even as late as the fourth century, there were ancient authorities, indeed some of the most ancient and weightiest, who either did not know of this utterance, so often quoted as characteristic of Jesus, or did not believe it had been uttered.

Many years ago, I received an anonymous letter, which abused me heartily for my want of moral courage in not speaking out. I thought that one of the oddest charges an anonymous letter-writer could bring. But I am not sure that the plentiful sowing of the pages of the article with which I am dealing with accusations of evasion, may not seem odder to those who consider that the main strength of the answers with which I have been favoured (in this review and elsewhere) is devoted, not to anything in the text of my first paper, but to a note which occurs at p. 212. In this I say :

Dr. Wace tells us : " It may be asked how far we can rely on the accounts we possess of our Lord's teaching on these subjects." And he seems to think the question appropriately answered by the assertion that it "ought to be regarded as settled by M. Renan's practical surrender of the adverse case."

I requested Dr. Wace to point out the passages of M. Renan's works in which, as he affirms, this

" practical surrender " (not merely as to the age
and authorship of the Gospels, be it observed, but
as to their historical value) is made, and he has
been so good as to do so. Now let us consider
the parts of Dr. Wace's citation from Renan which
are relevant to the issue :—

> The author of this Gospel [Luke] is certainly the same as the
> author of the Acts of the Apostles. Now the author of the
> Acts seems to be a companion of St. Paul—a character which
> accords completely with St. Luke.   I know that more than one
> objection may be opposed to this reasoning : but one thing, at
> all events, is beyond doubt, namely, that the author of the
> third Gospel and of the Acts is a man who belonged to the
> second apostolic generation ; and this suffices for our purpose.

This is a curious " practical surrender of the
adverse case."   M. Renan thinks that there is no
doubt that the author of the third Gospel is the
author of the Acts—a conclusion in which I
suppose critics generally agree. He goes on to
remark that this person *seems* to be a companion
of St. Paul, and adds that Luke was a companion
of St. Paul.   Then, somewhat needlessly, M.
Renan points out that there is more than one
objection to jumping, from such data as these, to
the conclusion that " Luke " is the writer of the
third Gospel.   And, finally, M. Renan is content
to reduce that which is " beyond doubt " to the
fact that the author of the two books is a man of
the second apostolic generation.   Well, it seems to
me that I could agree with all that M. Renan

considers "beyond doubt" here, without surren-
dering anything, either "practically" or theoretically.

Dr. Wace ("Nineteenth Century," March, p.
363) states that he derives the above citation
from the preface to the 15th edition of the "Vie
de Jésus." My copy of "Les Evangiles, dated
1877, contains a list of Renan's "Œuvres Com-
plètes," at the head of which I find "Vie de
Jésus," 15ᵉ édition. It is, therefore, a later work
than the edition of the "Vie de Jésus" which Dr.
Wace quotes. Now "Les Évangiles," as its name
implies, treats fully of the questions respecting
the date and authorship of the Gospels; and any
one who desired, not merely to use M. Renan's
expressions for controversial purposes, but to give
a fair account of his views in their full significance, would, I think, refer to the later source.

If this course had been taken, Dr. Wace might
have found some as decided expressions of opinion,
in favour of Luke's authorship of the third Gospel,
as he has discovered in "The Apostles." I mention this circumstance, because I desire to point
out that, taking even the strongest of Renan's
statements, I am still at a loss to see how it
justifies that large-sounding phrase, "practical
surrender of the adverse case." For, on p. 438 of
"Les Évangiles," Renan speaks of the way in
which Luke's "excellent intentions" have led him
to torture history in the Acts; he declares Luke

to be the founder of that " eternal fiction which is
called ecclesiastical history " ; and, on the preceding page, he talks of the "myth" of the
Ascension—with its " *mise en scène voulue.*" At
p. 435, I find " Luc, ou l'auteur quel qu'il soit du
troisième Évangile"; at p. 280, the accounts of
the Passion, the death and the resurrection of
Jesus, are said to be "peu historiques"; at p. 283,
" La valeur historique du troisième Évangile est
sûrement moindre que celles des deux premiers."
A Pyrrhic sort of victory for orthodoxy, this
" surrender"! And, all the while, the scientific
student of theology knows that, the more reason
there may be to believe that Luke was the companion of Paul, the more doubtful becomes his
credibility, if he really wrote the Acts. For, in
that case, he could not fail to have been acquainted
with Paul's account of the Jerusalem conference,
and he must have consciously misrepresented it.

We may next turn to the essential part of Dr.
Wace's citation ("Nineteenth Century," p. 365)
touching the first Gospel :—

St. Matthew evidently deserves peculiar confidence for
the discourses. Here are the "oracles"—the very notes taken
while the memory of the instruction of Jesus was living and
definite.

M. Renan here expresses the very general
opinion as to the existence of a collection of
" logia," having a different origin from the text

in which they are embedded, in Matthew.
"Notes" are somewhat suggestive of a shorthand
writer, but the suggestion is unintentional, for M.
Renan assumes that these "notes" were taken,
not at the time of the delivery of the "logia" but
subsequently, while (as he assumes) the memory
of them was living and definite; so that, in this
very citation, M. Renan leaves open the question
of the general historical value of the first Gospel;
while it is obvious that the accuracy of "notes"
taken, not at the time of delivery, but from
memory, is a matter about which more than one
opinion may be fairly held. Moreover, Renan
expressly calls attention to the difficulty of dis-
tinguishing the authentic "logia" from later
additions of the same kind ("Les Évangiles,"
p. 201). The fact is, there is no contradiction
here to that opinion about the first Gospel which
is expressed in " Les Évangiles " (p. 175).

The text of the so-called Matthew supposes the pre-existence
of that of Mark, and does little more than complete it. He
completes it in two fashions—first, by the insertion of those
long discourses which gave their chief value to the Hebrew
Gospels ; then by adding traditions of a more modern formation, results of successive developments of the legend, and to
which the Christian consciousness already attached infinite
value.

M. Renan goes on to suggest that besides
" Mark," " pseudo-Matthew " used an Aramaic
version of the Gospel, originally set forth in that

dialect. Finally, as to the second Gospel ("Nineteenth Century," p. 365) :—

> He [Mark] is full of minute observations, proceeding, beyond
> doubt, from an eye-witness. There is nothing to conflict with
> the supposition that this eye-witness . . . was the Apostle
> Peter himself, as Papias has it.

Let us consider this citation by the light of
" Les Évangiles " :—

> This work, although composed after the death of Peter, was,
> in a sense, the work of Peter ; it represents the way in which
> Peter was accustomed to relate the life of Jesus (p. 116).

M. Renan goes on to say that, as an historical
document, the Gospel of Mark has a great
superiority (p. 116) ; but Mark has a motive for
omitting the discourses, and he attaches a " puerile
importance " to miracles (p. 117). The Gospel of
Mark is less a legend, than a biography written
with credulity (p. 118). It would be rash to say
that Mark has not been interpolated and retouched (p. 120).

If any one thinks that I have not been warranted
in drawing a sharp distinction between " scientific
theologians " and " counsels for creeds " ; or that
my warning against the too ready acceptance of
certain declarations as to the state of biblical
criticism was needless ; or that my anxiety as to
the sense of the word " practical " was super-
fluous ; let him compare the statement that M.
Renan has made a " practical surrender of the

adverse case" with the facts just set forth. For
what is the adverse case? The question, as Dr.
Wace puts it, is, "It may be asked how far can
we rely on the accounts we possess of our Lord's
teaching on these subjects." It will be obvious
that M. Renan's statements amount to an adverse
answer—to a "practical" denial that any great
reliance can be placed on these accounts. He
does not believe that Matthew, the apostle, wrote
the first Gospel; he does not profess to know who
is responsible for the collection of "logia," or how
many of them are authentic; though he calls the
second Gospel the most historical, he points out
that it is written with credulity, and may have
been interpolated and retouched; and, as to the
author, "quel qu'il soit," of the third Gospel, who
is to "rely on the accounts" of a writer, who
deserves the cavalier treatment which "Luke"
meets with at M. Renan's hands?

I repeat what I have already more than once
said, that the question of the age and the authorship of the Gospels has not, in my judgment, the
importance which is so commonly assigned to it;
for the simple reason that the reports, even of
eye-witnesses, would not suffice to justify belief in
a large and essential part of their contents; on
the contrary, these reports would discredit the
witnesses. The Gadarene miracle, for example, is
so extremely improbable, that the fact of its being
reported by three, even independent, authorities

could not justify belief in it, unless we had the clearest evidence as to their capacity as observers and as interpreters of their observations. But it is evident that the three authorities are not independent ; that they have simply adopted a legend, of which there were two versions ; and instead of their proving its truth, it suggests their superstitious credulity : so that if " Matthew," " Mark," and " Luke " are really responsible for the Gospels, it is not the better for the Gadarene story, but the worse for them.

A wonderful amount of controversial capital has been made out of my assertion in the note to which I have referred, as an *obiter dictum* of no consequence to my argument, that if Renan's work[1] were non-extant, the main results of biblical criticism, as set forth in the works of Strauss, Baur, Reuss, and Volkmar, for example, would not be sensibly affected. I thought I had explained it satisfactorily already, but it seems that my explanation has only exhibited still more of my native perversity, so I ask for one more chance.

In the course of the historical development of any branch of science, what is universally observed is this : that the men who make epochs, and are the real architects of the fabric of exact knowledge, are those who introduce fruitful ideas or

---

[1] I trust it may not be supposed that I undervalue M. Renan's labours, or intended to speak slightingly of them.

theology, hereafter, to take that element into
serious consideration; so Baur, in giving prominence to the cardinal fact of the divergence of the
Nazarene and Pauline tendencies in the primitive
Church; so Reuss, in setting a marvellous example
of the cool and dispassionate application of the
principles of scientific criticism over the whole
field of Scripture; so Volkmar, in his clear and
forcible statement of the Nazarene limitations of
Jesus, contributed results of permanent value in
scientific theology. I took these names as they
occurred to me. Undoubtedly, I might have
advantageously added to them; perhaps, I might
have made a better selection. But it really is
absurd to try to make out that I did not know
that these writers widely disagree; and I believe
that no scientific theologian will deny that, in
principle, what I have said is perfectly correct.
Ecclesiastical advocates, of course, cannot be
expected to take this view of the matter. To
them, these mere seekers after truth, in so far as
their results are unfavourable to the creed the
clerics have to support, are more or less " infidels,"
or favourers of " infidelity"; and the only thing
they care to see, or probably can see, is the fact
that, in a great many matters, the truth-seekers
differ from one another, and therefore can easily
be exhibited to the public, as if they did nothing
else; as if any one who referred to their having,
each and all, contributed his share to the results

debarred from freely following out scientific methods to their legitimate conclusions, whatever those conclusions may be. If I may borrow a phrase paraded at the Church Congress, I think it " ought to be unpleasant " for any man of science to find himself in the position of such a teacher.

Human nature is not altered by seating it in a professorial chair, even of theology.   I have very little doubt that if, in the year 1859, the tenure of my office had depended upon my adherence to the doctrines of Cuvier, the objections to them set forth in the " Origin of Species " would have had a halo of gravity about them that, being free to teach what I pleased, I failed to discover.   And, in making that statement, it does not appear to me that I am confessing that I should have been debarred by " selfish interests " from making candid inquiry, or that I should have been biassed by " sordid motives."   I hope that even such a fragment of moral sense as may remain in an ecclesiastical "infidel" might have got me through the difficulty; but it would be unworthy to deny, or disguise, the fact that a very serious difficulty must have been created for me by the nature of my tenure.   And let it be observed that the temptation, in my case, would have been far slighter than in that of a professor of theology; whatever biological doctrine I had repudiated, nobody I cared for would have thought the worse of me for so doing.   No scientific journals would have howled me down, as the religious newspapers howled down my too honest friend, the late Bishop of Natal; nor would my colleagues of the Royal Society have turned their backs upon me, as his episcopal colleagues boycotted him.

I say these facts are obvious, and that it is wholesome and needful that they should be stated. It is in the interests of theology, if it be a science, and it is in the interests of those teachers of theology who desire to be something better than counsel for creeds, that it should be taken to heart. The seeker after theological truth and that only, will no more suppose that I have insulted him, than the prisoner who works in fetters will try to pick a quarrel with me, if I suggest that he would get on better if the fetters were knocked off; unless indeed, as it is said does happen in the course of long captivities, that the victim at length ceases to feel the weight of his chains, or even takes to hugging them, as if they were honourable ornaments.[1]

[1] To-day's *Times* contains a report of a remarkable speech by Prince Bismarck, in which he tells the Reichstag that he has long given up investing in foreign stock, lest so doing should mislead his judgment in his transactions with foreign states. Does this declaration prove that the Chancellor accuses himself of being "sordid" and "selfish"; or does it not rather show that, even in dealing with himself, he remains the man of realities?

X

THE KEEPERS OF THE HERD OF SWINE

[1890]

I HAD fondly hoped that Mr. Gladstone and I had come to an end of disputation, and that the hatchet of war was finally superseded by the calumet, which, as Mr. Gladstone, I believe, objects to tobacco, I was quite willing to smoke for both. But I have had, once again, to discover that the adage that whoso seeks peace will ensue it, is a somewhat hasty generalisation. The renowned warrior with whom it is my misfortune to be opposed in most things has dug up the axe and is on the war-path once more. The weapon has been wielded with all the dexterity which long practice has conferred on a past master in craft, whether of wood or state. And I have reason to believe that the simpler sort of the great tribe which he heads, imagine that my scalp is already on its way to adorn their big chief's wigwam. I am glad therefore to be able to

relieve any anxieties which my friends may
entertain without delay. I assure them that my
skull retains its normal covering, and that though,
naturally, I may have felt alarmed, nothing
serious has happened. My doughty adversary
has merely performed a war dance, and his blows
have for the most part cut the air. I regret to
add, however, that by misadventure, and I am
afraid I must say carelessness, he has inflicted
one or two severe contusions on himself.

When the noise of approaching battle roused
me from the dreams of peace which occupy my
retirement, I was glad to observe (since I must
fight) that the campaign was to be opened upon
a new field. When the contest raged over the
Pentateuchal myth of the creation, Mr. Gladstone's
manifest want of acquaintance with the facts and
principles involved in the discussion, no less than
with the best literature on his own side of the
subject, gave me the uncomfortable feeling that I
had my adversary at a disadvantage. The sun of
science, at my back, was in his eyes. But, on the
present occasion, we are happily on an equality.
History and Biblical criticism are as much, or
as little, my vocation as they are that of Mr.
Gladstone; the blinding from too much light, or
the blindness from too little, may be presumed to
be equally shared by both of us.

Mr. Gladstone takes up his new position in the
country of the Gadarenes. His strategic sense

justly leads him to see that the authority of the
teachings of the synoptic Gospels, touching the
nature of the spiritual world, turns upon the
acceptance, or the rejection, of the Gadarene and
other like stories.   As we accept, or repudiate,
such histories as that of the possessed pigs, so
shall we accept, or reject, the witness of the
synoptics to such miraculous interventions.

It is exactly because these stories constitute
the key-stone of the orthodox arch, that I
originally drew attention to them; and, in spite
of my longing for peace, I am truly obliged to
Mr. Gladstone for compelling me to place my case
before the public once more.   It may be thought
that this is a work of supererogation by those
who are aware that my essay is the subject of
attack in a work so largely circulated as the
"Impregnable Rock of Holy Scripture"; and who
may possibly, in their simplicity, assume that it
must be truthfully set forth in that work.   But
the warmest admirers of Mr. Gladstone will hardly
be prepared to maintain that mathematical accuracy in stating the opinions of an opponent is the
most prominent feature of his controversial method.
And what follows will show that, in the present
case, the desire to be fair and accurate, the
existence of which I am bound to assume, has
not borne as much fruit as might have been
expected.

In referring to the statement of the narrators,

that the herd of swine perished in consequence of the entrance into them of the demons by the permission, or order, of Jesus of Nazareth, I said:

"Everything that I know of law and justice convinces me that the wanton destruction of other people's property is a misdemeanour of evil example" ("Nineteenth Century," February, 1889, p. 172).

Mr. Gladstone has not found it convenient to cite this passage; and, in view of various considerations, I dare not assume that he would assent to it, without sundry subtle modifications which, for me, might possibly rob it of its argumentative value. But, until the proposition is seriously controverted, I shall assume it to be true, and content myself with warning the reader that neither he nor I have any grounds for assuming Mr. Gladstone's concurrence. With this caution, I proceed to remark that I think it may be granted that the people whose herd of 2000 swine (more or fewer) was suddenly destroyed suffered great loss and damage. And it is quite certain that the narrators of the Gadarene story do not, in any way, refer to the point of morality and legality thus raised; as I said, they show no inkling of the moral and legal difficulties which arise.

Such being the facts of the case, I submit that for those who admit the principle laid down, the conclusion which I have drawn necessarily follows;

though I repeat that, since Mr. Gladstone does
not explicitly admit the principle, I am far from
suggesting that he is bound by its logical consequences.   However, I distinctly reiterate the
opinion that any one who acted in the way
described in the story would, in my judgment,
be guilty of "a misdemeanour of evil example."
About that point I desire to leave no ambiguity
whatever; and it follows that, if I believed the
story, I should have no hesitation in applying
this judgment to the chief actor in it.

But, if any one will do me the favour to turn
to the paper in which these passages occur, he
will find that a considerable part of it is devoted
to the exposure of the familiar trick of the
"counsel for creeds," who, when they wish to
profit by the easily stirred *odium theologicum*, are
careful to confuse disbelief in a narrative of a
man's act, or disapproval of the acts as narrated,
with disbelieving and vilipending the man himself.
If I say that "according to paragraphs in several
newspapers, my valued Separatist friend A. B. has
houghed a lot of cattle, which he considered to be
unlawfully in the possession of an Irish land-
grabber; that, in my opinion, any such act is a
misdemeanour of evil example; but, that I utterly
disbelieve the whole story and have no doubt that
it is a mere fabrication:" it really appears to me
that, if any one charges me with calling A. B. an
immoral misdemeanant, I should be justified in

using very strong language respecting either his sanity or his veracity. And, if an analogous charge has been brought in reference to the Gadarene story, there is certainly no excuse producible, on account of any lack of plain speech on my part. Surely no language can be more explicit than that which follows :

"I can discern no escape from this dilemma; either Jesus said what he is reported to have said, or he did not. In the former case, it is inevitable that his authority on matters connected with the 'unseen world' should be roughly shaken; in the latter, the blow falls upon the authority of the synoptic Gospels" (p. 173). "The choice then lies between discrediting those who compiled the Gospel biographies and disbelieving the Master, whom they, simple souls, thought to honour by preserving such traditions of the exercise of his authority over Satan's invisible world" (p. 174)· And I leave no shadow of doubt as to my own choice: "After what has been said, I do not think that any sensible man, unless he happen to be angry, will accuse me of 'contradicting the Lord and his Apostles' if I reiterate my total disbelief in the whole Gadarene story" (p. 178).

I am afraid, therefore, that Mr. Gladstone must have been exceedingly angry when he committed himself to such a statement as follows :

So, then, after eighteen centuries of worship offered to our Lord by the most cultivated, the most developed, and the most

progressive portion of the human race, it has been reserved to a
scientific inquirer to discover that He was no better than a law-
breaker and an evil-doer. . . . How, in such a matter, came the
honours of originality to be reserved to our time and to Professor
Huxley? (Pp. 269, 270.)

Truly, the hatchet is hardly a weapon of precision, but would seem to have rather more the
character of the boomerang, which returns to
damage the reckless thrower.   Doubtless such
incidents are somewhat ludicrous.   But they have
a very serious side ; and, if I rated the opinion of
those who blindly follow Mr. Gladstone's leading,
but not light, in these matters, much higher than
the great Duke of Wellington's famous standard
of minimum value, I think I might fairly beg
them to reflect upon the general bearings of this
particular example of his controversial method.
I imagine it can hardly commend itself to their
cool judgment.

After this tragi-comical ending to what an old
historian calls a "robustious and rough coming
on"; and after some praises of the provisions of
the Mosaic law in the matter of not eating pork—
in which, as pork disagrees with me and for some
other reasons, I am much disposed to concur,
though I do not see what they have to do with
the matter in hand—comes the serious onslaught.

Mr. Huxley, exercising his rapid judgment on the text, does
not appear to have encumbered himself with the labour of inquiring what anybody else had known or said about it.   He has

thus missed a point which might have been set up in support of his accusation against our Lord. (P. 273.)

Unhappily for my comfort, I have been much exercised in controversy during the past thirty years ; and the only compensation for the loss of time and the trials of temper which it has inflicted upon me, is that I have come to regard it as a branch of the fine arts, and to take an impartial and æsthetic interest in the way in which it is conducted, even by those whose efforts are directed against myself. Now, from the purely artistic point of view (which, as we are all being told, has nothing to do with morals), I consider it an axiom, that one should never appear to doubt that the other side has performed the elementary duty of acquiring proper elementary information, unless there is demonstrative evidence to the contrary. And I think, though I admit that this may be a purely subjective appreciation, that (unless you are quite certain) there is a " want of finish," as a great master of disputation once put it, about the suggestion that your opponent has missed a point on his own side. Because it may happen that he has not missed it at all, but only thought it unworthy of serious notice. And if he proves that, the suggestion looks foolish.

Merely noting the careful repetition of a charge, the absurdity of which has been sufficiently exposed above, I now ask my readers to accompany me on a little voyage of discovery in search of the side on which the rapid ju^dgment and the ignorance of the literature of the subject lie. I think I may promise them very little trouble, and a good deal of entertainment.

Mr. Gladstone is of opinion that the Gadarene swinefolk were " Hebrews bound by the Mosaic law " (p. 274) ; and he conceives that it has not occurred to me to learn what may be said in favour of and against this view. He tells us that

> Some commentators have alleged the authority of Josephus for stating that Gadara was a city of Greeks rather than of Jews, from whence it might be inferred that to keep swine was inno-cent and lawful. (P. 273.)

Mr. Gladstone then goes on to inform his readers that in his painstaking search after truth he has submitted to the labour of personally examining the writings of Josephus. Moreover, in a note, he positively exhibits an acquaintance, in addition, with the works of Bishop Wordsworth and of Archbishop Trench ; and even shows that he has read Hudson's commentary on Josephus. And yet people say that our Biblical critics do not equal the Germans in research ! But Mr. Gladstone s citation of Cuvier and Sir John Herschel about the Creation myth, and his ignor-ance of all the best modern writings on his own side, produced a great impression on my mind. I have had the audacity to suspect that his ac-quaintance with what has been done in Biblical history might stand at no higher level than his information about the natural sciences. However unwillingly, I have felt bound to consider the possibility that Mr. Gladstone's labours in this matter may have carried him no further than Josephus and the worthy, but somewhat antique, episcopal and other authorities to whom he refers; that even his reading of Josephus may have been of the most cursory nature, directed not to the understanding of his author, but to the discovery of useful controversial matter; and that, in view of the not inconsiderable misrepresentation of my statements to which I have drawn attention, it might be that Mr. Gladstone's exposition of the evidence of Josephus was not more trustworthy. I proceed to show that my previsions have been fully justified. I doubt if controversial literature contains anything more *piquant* than the story I have to unfold.

That I should be reproved for rapidity of judgment is very just: however quaint the situation of Mr. Gladstone, as the reprover, may seem to people blessed with a sense of humour. But it is a quality, the defects of which have been painfully obvious to me all my life; and I try to keep my Pegasus—at best, a poor Shetland variety of that species of quadruped—at a respectable jog-trot, by loading him heavily with bales of reading. Those who took the trouble to study my paper in good faith and not for mere controversial purposes,

have a right to know, that something more than a
hasty glimpse of two or three passages of Josephus
(even with as many episcopal works thrown in)
lay at the back of the few paragraphs I devoted to
the Gadarene story. I proceed to set forth, as
briefly as I can, some results of that preparatory
work. My artistic principles do not permit me, at
present, to express a doubt that Mr. Gladstone
was acquainted with the facts I am about to
mention when he undertook to write. But, if he
did know them, then both what he has said and
what he has not said, his assertions and his
omissions alike, will require a paragraph to themselves.

The common consent of the synoptic Gospels
affirms that the miraculous transference of devils
from a man, or men, to sundry pigs, took place
somewhere on the eastern shore of the Lake of
Tiberias ; " on the other side of the sea over
against Galilee," the western shore being, without
doubt, included in the latter province. But there
is no such concord when we come to the name of
the part of the eastern shore, on which, according
to the story, Jesus and his disciples landed. In the
revised version, Matthew calls it the " country of
the Gadarenes : " Luke and Mark have " Gerasenes."
In sundry very ancient manuscripts " Gergesenes "
occurs.

The existence of any place called Gergesa, however, is declared by the weightiest authorities

whom I have consulted to be very questionable;
and no such town is mentioned in the list of the
cities of the Decapolis, in the territory of which
(as it would seem from Mark v. 20) the transaction
was supposed to take place. About Gerasa, on
the other hand, there hangs no such doubt. It
was a large and important member of the group
of the Decapolitan cities. But Gerasa is more than
thirty miles distant from the nearest part of the
Lake of Tiberias, while the city mentioned in the
narative could not have been very far off the scene
of the event. However, as Gerasa was a very important Hellenic city, not much more than a score
of miles from Gadara, it is easily imaginable that
a locality which was part of Decapolitan territory
may have been spoken of as belonging to one of
the two cities, when it really appertained to the
other. After weighing all the arguments, no
doubt remains on my mind that "Gadarene"
is the proper reading. At the period under consideration, Gadara appears to have been a good-
sized fortified town, about two miles in circumference. It was a place of considerable strategic
importance, inasmuch as it lay on a high ridge at
the point of intersection of the roads from Tiberias,
Scythopolis, Damascus, and Gerasa. Three miles
north from it, where the Tiberias road descended
into the valley of the Hieromices, lay the famous
hot springs and the fashionable baths of Amatha.
On the north-east side, the remains of the extensive

necropolis of Gadara are still to be seen. Innumerable sepulchral chambers are excavated in the limestone cliffs, and many of them still contain sarcophaguses of basalt; while not a few are converted into dwellings by the inhabitants of the present village of Um Keis. The distance of Gadara from the south-eastern shore of the Lake of Tiberias is less than seven miles. The nearest of the other cities of the Decapolis, to the north, is Hippos, which also lay some seven miles off, in the south-eastern corner of the shore of the lake. In accordance with the ancient Hellenic practice, that each city should be surrounded by a certain amount of territory amenable to its jurisdiction,[1] and on other grounds, it may be taken for certain that the intermediate country was divided between Gadara and Hippos; and that the citizens of Gadara had free access to a port on the lake. Hence the title of "country of the Gadarenes" applied to the locality of the porcine catastrophe becomes easily intelligible. The swine may well be imagined to have been feeding (as they do now in the adjacent region) on the hillsides, which slope somewhat steeply down to the lake from the northern boundary wall of the valley of the Hieromices (*Nahr Yarmuk*), about half-way between the city

[1] Thus Josephus (lib. ix.) says that his rival, Justus, persuaded the citizens of Tiberias to "set the villages that belonged to Gadara and Hippos on fire; which villages were situated on the borders of Tiberias and of the region of Scythopolis."

and the shore, and doubtless lay well within the
territory of the *polis* of Gadara.

The proof that Gadara was, to all intents and purposes, a Gentile, and not a Jewish, city is complete.
The date and the occasion of its foundation are
unknown; but it certainly existed in the third
century B.C. Antiochus the Great annexed it to
his dominions in B.C. 198. After this, during
the brief revival of Jewish autonomy, Alexander
Jannæus took it; and for the first time, so far as
the records go, it fell under Jewish rule.[1] From
this it was rescued by Pompey (B.C. 63), who
rebuilt the city and incorporated it with the
province of Syria. In gratitude to the Romans
for the dissolution of a hated union, the Gadarenes
adopted the Pompeian era on their coinage.
Gadara was a commercial centre of some importance, and therefore, it may be assumed, Jews
settled in it, as they settled in almost all considerable Gentile cities. But a wholly mistaken
estimate of the magnitude of the Jewish colony
has been based upon the notion that Gabinius,
proconsul of Syria in 57-55 B.C., seated one of the
five sanhedrims in Gadara. Schürer has pointed
out that what he really did was to lodge one of
them in Gazara, far away on the other side of the
Jordan. This is one of the many errors which have
arisen out of the confusion of the names Ga*d*ara,
Ga*z*ara, and Ga*b*ara.

[1] It is said to have been destroyed by its captors.

Augustus made a present of Gadara to Herod the Great, as an appanage personal to himself; and, upon Herod's death, recognising it to be a " Grecian city " like Hippos and Gaza,[1] he transferred it back to its former place in the province of Syria. That Herod made no effort to judaise his temporary possession, but rather the contrary, is obvious from the fact that the coins of Gadara, while under his rule, bear the image of Augustus with the superscription $\Sigma\epsilon\beta\alpha\sigma\tau\sigma$—a flying in the face of Jewish prejudices which, even he, did not dare to venture upon in Judæa. And I may remark that, if my co-trustee of the British Museum had taken the trouble to visit the splendid numismatic collection under our charge, he might have seen two coins of Gadara, one of the time of Tiberius and the other of that of Titus, each bearing the effigies of the emperor on the obverse : while the personified genius of the city is on the reverse of the former. Further, the well-known works of De Saulcy and of Ekhel would have supplied the information that, from the time of Augustus to that of Gordian, the Gadarene coinage had the same thoroughly Gentile character. Curious that a city of " Hebrews bound by the Mosaic law " should tolerate such a mint !

---

[1] " But as to the Grecian cities, Gaza and Gadara and Hippos, he cut them off from the kingdom and added them to Syria." —Josephus, *Wars*, II. vi. 3. See also *Antiquities*, XVII. xi. 4.

Whatever increase in population the Ghetto of Gadara may have undergone, between B.C. 4 and A.D. 66, it nowise affected the gentile and anti-judaic character of the city at the outbreak of the great war ; for Josephus tells us that, immediately after the great massacre of Cæsarea, the revolted Jews "laid waste the villages of the Syrians and their neighbouring cities, Philadelphia and Se-bonitis and Gerasa and Pella and Scythopolis, and after them Gadara and Hippos " (" Wars," II. xviii. 1).   I submit that, if Gadara had been a city of "Hebrews bound by the Mosaic law," the ravaging of their territory by their brother Jews, in revenge for the massacre of the Cæsarean Jews by the Gentile population of that place, would surely have been a somewhat unaccountable pro-ceeding.   But when we proceed a little further, to the fifth section of the chapter in which this state-ment occurs, the whole affair becomes intelligible enough.

Besides this murder at Scythopolis, the other cities rose up against the Jews that were among them : those of Askelon slew two thousand five hundred, and those of Ptolemais two thousand, and put not a few into bonds ; those of Tyre also put a great number to death, but kept a greater number in prison ; more-over, those of Hippos and those of Gadara did the like, while they put to death the boldest of the Jews, but kept those of whom they were most afraid in custody ; as did the rest of the cities of Syria according as they every one either hated them or were afraid of them.

Josephus is not always trustworthy, but he has

no conceivable motive for altering facts here; he speaks of contemporary events, in which he himself took an active part, and he characterises the cities in the way familiar to him. For Josephus, Gadara is just as much a Gentile city as Ptolemais; it was reserved for his latest commentator, either ignoring, or ignorant of, all this, to tell us that Gadara had a Hebrew population, bound by the Mosaic law.

In the face of all this evidence, most of which has been put before serious students, with full reference to the needful authorities and in a thoroughly judicial manner, by Schürer in his classical work,[1] one reads with stupefaction the statement which Mr. Gladstone has thought fit to put before the uninstructed public:

> Some commentators have alleged the authority of Josephus for stating that Gadara was a city of Greeks rather than of Jews, from whence it might be inferred that to keep swine was innocent and lawful. This is not quite the place for a critical examination of the matter; but I have examined it, and have satisfied myself that Josephus gives no reason whatever to suppose that the population of Gadara, and still less (if less may be) the population of the neighbourhood, and least of all the swine-herding or lower portion of that population, were other than Hebrews bound by the Mosaic law. (Pp. 373-4.)

Even "rapid judgment" cannot be pleaded in excuse for this surprising statement, because a "Note on the Gadarene miracle" is added (in a special appendix), in which the references are

---

[1] *Geschichte des jüdischen Volkes im Zeitalter Christi*, 1886-90.

given to the passages of Josephus, by the improved interpretation of which, Mr. Gladstone
has thus contrived to satisfy himself of the thing
which is not. One of these is "Antiquities" XVII.
xiii. 4, in which section, I regret to say, I can find
no mention of Gadara. In "Antiquities," XVII. xi.
4, however, there is a passage which would appear
to be that which Mr. Gladstone means; and I will
give it in full, although I have already cited part
of it :

There were also certain of the cities which paid tribute to
Archelaus ; Strato's tower, and Sebaste, with Joppa and Jerusalem ; for, as to Gaza, Gadara, and Hippos, they were Grecian
cities, which Cæsar separated from his government, and added
them to the province of Syria.

That is to say, Augustus simply restored the state
of things which existed before he gave Gadara,
then certainly a Gentile city, lying outside Judæa,
to Herod as a mark of great personal favour. Yet
Mr. Gladstone can gravely tell those who are not
in a position to check his statements :

The sense seems to be, not that these cities were inhabited by
a Greek population, but that they had politically been taken out
of Judæa and added to Syria, which I presume was classified as
simply Hellenic, a portion of the great Greek empire erected by
Alexander. (Pp. 295-6.)

Mr. Gladstone's next reference is to the " Wars,"
III. vii. 1 :

So Vespasian marched to the city Gadara, and took it upon
the first onset, because he found it destitute of a considerable

number of men grown up fit for war. He then came into it, and slew all the youth, the Romans having no mercy on any age whatsoever; and this was done out of the hatred they bore the nation, and because of the iniquity they had been guilty of in the affair of Cestius.

Obviously, then, Gadara was an ultra-Jewish city. Q.E.D. But a student trained in the use of weapons of precision, rather than in that of rhetorical tomahawks, has had many and painful warnings to look well about him, before trusting an argument to the mercies of a passage, the context of which he has not carefully considered. If Mr. Gladstone had not been too much in a hurry to turn his imaginary prize to account—if he had paused just to look at the preceding chapter of Josephus—he would have discovered that his much haste meant very little speed. He would have found ("Wars," III. vi. 2) that Vespasian marched from his base, the port of Ptolemais (Acre), on the shores of the Mediterranean, into Galilee; and, having dealt with the so-called "Gadara," was minded to finish with Jotapata, a strong place about fourteen miles south-east of Ptolemais, into which Josephus, who at first had fled to Tiberias, eventually threw himself— Vespasian arriving before Jotapata "the very next day." Now, if any one will take a decent map of Ancient Palestine in hand, he will see that Jotapata, as I have said, lies about fourteen miles in a straight line east-south-east of Ptolemais,

while a certain town, " Gabara " (which was also
held by the Jews), is situated, about the same
distance, to the east of that port. Nothing can be
more obvious than that Vespasian, wishing to
advance from. Ptolemais into Galilee, could not
afford to leave these strongholds in the possession
of the enemy; and, as Gabara would lie on his
left flank when he moved to Jotapata, he took
that city, whence his communications with his
base could easily be threatened, first. It might
really have been fair evidence of demoniac possession, if the best general of Rome had marched
forty odd miles, as the crow flies, through hostile
Galilee, to take a city (which, moreover, had just
tried to abolish its Jewish population) on the
other side of the Jordan; and then marched back
again to a place fourteen miles off his starting-
point.[1] One would think that the most careless
of readers must be startled by this incongruity
into inquiring whether there might not be something wrong with the text; and, if he had done so,
he would have easily discovered that since the
time of Reland, a century and a half ago, careful
scholars have read Gabara for Gadara.[1]

Once more, I venture to point out that training

[1] If William the Conqueror, after fighting the battle of
Hastings, had marched to capture Chichester and then returned
to assault Rye, being all the while anxious to reach London, his
proceedings would not have been more eccentric than Mr. Gladstone must imagine those of Vespasian were.

[2] See Reland, *Palestina* (1714), t. ii. p. 771. Also Robinson,
Later Biblical Researches (1856), p. 87 *note*.

in the use of the weapons of precision of science
may have its value in historical studies, if only in
preventing the occurrence of droll blunders in
geography.

In the third citation ("Wars," IV. vii.) Josephus
tells us that Vespasian marched against "Gadara,"
which he calls the metropolis of Peræa (it was
possibly the seat of a common festival of the
Decapolitan cities), and entered it, without opposition, the wealthy and powerful citizens having
opened negotiations with him without the knowledge of an opposite party, who, "as being inferior
in number to their enemies, who were within the
city, and seeing the Romans very near the city,"
resolved to fly.   Before doing so, however, they,
after a fashion unfortunately too common among
the Zealots, murdered and shockingly mutilated
Dolesus, a man of the first rank, who had pro-
moted the embassy to Vespasian; and then "ran
out of the city." Hereupon, "the people of
Gadara" (surely not this time "Hebrews bound
by the Mosaic law") received Vespasian with joy-
ful acclamations, voluntarily pulled down their
wall, so that the city could not in future be used
as a fortress by the Jews, and accepted a Roman
garrison for their future protection. Granting
that this Gadara really is the city of the
Gadarenes, the reference, without citation, to the
passage, in support of Mr. Gladstone's contention
seems rather remarkable. Taken in conjunction

with the shortly antecedent ravaging of the Gadarene territory by the Jews, in fact, better proof could hardly be expected of the real state of the case; namely, that the population of Gadara (and notably the wealthy and respectable part of it) was thoroughly Hellenic; though, as in Cæsarea and elsewhere among the Palestinian cities, the rabble contained a considerable body of fanatical Jews, whose reckless ferocity made them, even though a mere minority of the population, a standing danger to the city.

Thus Mr. Gladstone's conclusion from his study of Josephus, that the population of Gadara were "Hebrews bound by the Mosaic law," turns out to depend upon nothing better than a marvellously complete misinterpretation of what that author says, combined with equally marvellous geographical misunderstandings, long since exposed and rectified; while the positive evidence that Gadara, like other cities of the Decapolis, was thoroughly Hellenic in organisation, and essentially Gentile in population, is overwhelming.

And, that being the fact of the matter, patent to all who will take the trouble to enquire about what has been said about it, however obscure to those who merely talk of so doing, the thesis that the Gadarene swineherds, or owners, were Jews violating the Mosaic law shows itself to be an empty and most unfortunate guess. But really, whether they that kept the swine were Jews, or

c c 2

whether they were Gentiles, is a consideration which has no relevance whatever to my case. The legal provisions, which alone had authority over an inhabitant of the country of the Gadarenes, were the Gentile laws sanctioned by the Roman suzerain of the province of Syria, just as the only law, which has authority in England, is that re- cognised by the sovereign Legislature. Jewish communities in England may have their private code, as they doubtless had in Gadara. But an English magistrate, if called upon to enforce their peculiar laws, would dismiss the complainants from the judgment seat, let us hope with more politeness than Gallio did in a like case, but quite as firmly. Moreover, in the matter of keeping pigs, we may be quite certain that Gadarene law left everybody free to do as he pleased indeed encouraged the practice rather than otherwise. Not only was pork one of the commonest and one of the most favourite articles of Roman diet; but, to both Greeks and Romans, the pig was a sacri- ficial animal of high importance. Sucking pigs played an important part in Hellenic purificatory rites; and everybody knows the significance of the Roman suovetaurilia, depicted on so many bas- reliefs.

Under these circumstances, only the extreme need of a despairing "reconciler" drowning in a sea of adverse facts, can explain the catching at such a poor straw as the reckless guess that the

swineherds of the "country of the Gadarenes"
were erring Jews, doing a little clandestine business on their own account. The endeavour to
justify the asserted destruction of the swine by the
analogy of breaking open a cask of smuggled
spirits, and wasting their contents on the ground,
is curiously unfortunate. Does Mr. Gladstone
mean to suggest that a Frenchman landing at
Dover, and coming upon a cask of smuggled brandy
in the course of a stroll along the cliffs, has the
right to break it open and waste its contents on
the ground? Yet the party of Galileans who,
according to the narrative, landed and took a walk
on the Gadarene territory, were as much foreigners
in the Decapolis as Frenchmen would be at Dover.
Herod Antipas, their sovereign, had no jurisdiction in the Decapolis—they were strangers and
aliens, with no more right to interfere with a pig-
keeping Hebrew, than I have a right to interfere
with an English professor of the Israelitic faith, if
I see a slice of ham on his plate. According to
the law of the country in which these Galilean
foreigners found themselves, men might keep pigs
if they pleased. If the men who kept them were
Jews, it might be permissible for the strangers to
inform the religious authority acknowledged by the
Jews of Gadara; but to interfere themselves, in such
a matter, was a ep devoid of either moral or legal
justification.

Suppose a modern English Sabbatarian fanatic,

who believes, on the strength of his interpretation of the fourth commandment, that it is a deadly sin to work on the "Lord's Day," sees a fellow Puritan yielding to the temptation of getting in his harvest on a fine Sunday morning—is the former justified in setting fire to the latter's corn? Would not an English court of justice speedily teach him better?

In truth, the government which permits private persons, on any pretext (especially pious and patriotic pretexts), to take the law into their own hands, fails in the performance of the primary duties of all governments; while those who set the example of such acts, or who approve them, or who fail to disapprove them, are doing their best to dissolve civil society: they are compassers of illegality and fautors of immorality.

I fully understand that Mr. Gladstone may not see the matter in this light. He may possibly consider that the union of Gadara with the Decapolis, by Augustus, was a "blackguard" transaction, which deprived Hellenic Gadarene law of all moral force; and that it was quite proper for a Jewish Galilean, going back to the time when the land of the Girgashites was given to his ancestors, some 1500 years before, to act, as if the state of things which ought to obtain, in territory which traditionally, at any rate, belonged to his fore-fathers, did really exist. And, that being so, I can only say I do not agree with him, but leave the matter to the appreciation of those of our countrymen, happily not yet the minority, who believe that the first condition of enduring liberty is obedience to the law of the land.

The end of the month drawing nigh, I thought it well to send away the manuscript of the foregoing pages yesterday, leaving open, in my own mind, the possibility of adding a succinct characterisation of Mr. Gladstone's controversial methods as illustrated therein. This morning, however, I had the pleasure of reading a speech which I think must satisfy the requirements of the most fastidious of controversial artists ; and there occurs in it so concise, yet so complete, a delineation of Mr. Gladstone's way of dealing with disputed questions of another kind, that no poor effort of mine could better it as a description of the aspect which his treatment of scientific, historical, and critical questions presents to me.

> The smallest examination would have told a man of his capacity and of his experience that he was uttering the grossest exaggerations, that he was basing arguments upon the slightest hypotheses, and that his discussions only had to be critically examined by the most careless critic in order to show their intrinsic hollowness.

Those who have followed me through this paper will hardly dispute the justice of this judgment, severe as it is. But the Chief Secretary for

Ireland has science in the blood; and has the advantage of a natural, as well as a highly cultivated, aptitude for the use of methods of precision in investigation, and for the exact enunciation of the results thereby obtained.

XI

ILLUSTRATIONS OF MR. GLADSTONE'S CONTROVERSIAL METHODS

[1891]

THE series of essays, in defence of the historical accuracy of the Jewish and Christian Scriptures, contributed by Mr. Gladstone to "Good Words," having been revised and enlarged by their author, appeared last year as a separate volume, under the somewhat defiant title of "The Impregnable Rock of Holy Scripture."

The last of these Essays, entitled "Conclusion," contains an attack, or rather several attacks, couched in language which certainly does not err upon the side of moderation or of courtesy, upon statements and opinions of mine. One of these assaults is a deliberately devised attempt, not merely to rouse the theological prejudices ingrained in the majority of Mr. Gladsone's readers, but to hold me up as a person who has endeavoured to besmirch the personal character of the object of their veneration. For Mr. Gladstone asserts that

I have undertaken to try "the character of our
Lord" (p. 268); and he tells the many who are, as
I think unfortunately, predisposed to place implicit credit in his assertions, that it has been
reserved for me to discover that Jesus "was no
better than a law-breaker and an evil-doer!"
(p. 269).

It was extremely easy for me to prove, as I did
in the pages of this Review last December, that,
under the most favourable interpretation, this
amazing declaration must be ascribed to extreme
confusion of thought. And, by bringing an
abundance of good-will to the consideration of the
subject, I have now convinced myself that it is
right for me to admit that a person of Mr. Gladstone's intellectual acuteness really did mistake
the reprobation of the course of conduct ascribed
to Jesus, in a story of which I expressly say I do
not believe a word, for an attack on his character
and a declaration that he was "no better than a
law-breaker, and an evil-doer." At any rate, so far
as I can see, this is what Mr. Gladstone wished
to be believed when he wrote the following
passage :—

I must, however, in passing, make the confession that I did
not state with accuracy, as I ought to have done, the precise
form of the accusation. I treated it as an imputation on the
action of our Lord ; he replies that it is only an imputation on
the narrative of three evangelists respecting Him. The difference, from his point of view, is probably material, and I therefore regret that I overlooked it.[1]

---

[1] *Nineteenth Century*, February 1891, pp. 339-40.

Considering the gravity of the error which is here admitted, the fashion of the withdrawal appears more singular than admirable. From my "point of view"—not from Mr. Gladstone's apparently—the little discrepancy between the facts and Mr. Gladstone's carefully offensive travesty of them is "probably" (only "probably") material. However, as Mr Gladstone concludes with an official expression of regret for his error, it is my business to return an equally official expression of gratitude for the attenuated reparation with which I am favoured.

Having cleared this specimen of Mr. Gladstone's controversial method out of the way, I may proceed to the next assault, that on a passage in an article on Agnosticism ("Nineteenth Century," February 1889), published two years ago. I there said, in referring to the Gadarene story, "Everything I know of law and justice convinces me that the wanton destruction of other people's property is a misdemeanour of evil example." On this, Mr. Gladstone, continuing his candid and urbane observations, remarks ("Impregnable Rock," p. 273) that, "Exercising his rapid judgment on the text," and "not inquiring what anybody else had known or said about it," I had missed a point in support of that "accusation against our Lord" which he has now been constrained to admit I never made.

The "point" in question is that "Gadara was a

city of Greeks rather than of Jews, from whence
it might be inferred that to keep swine was
innocent and lawful." I conceive that I have
abundantly proved that Gadara answered exactly
to the description here given of it; and I shall
show, by and by, that Mr. Gladstone has used
language which, to my mind, involves the admission
that the authorities of the city were not Jews.
But I have also taken a good deal of pains to
show that the question thus raised is of no
importance in relation to the main issue.[1] If
Gadara was, as I maintain it was, a city of the
Decapolis, Hellenistic in constitution and containing a predominantly Gentile population, my
case is superabundantly fortified. On the other
hand, if the hypothesis that Gadara was under
Jewish government, which Mr. Gladstone seems
sometimes to defend and sometimes to give up,
were accepted, my case would be nowise weakened.
At any rate, Gadara was not included within the
jurisdiction of the tetrach of Galilee; if it had
been, the Galileans who crossed over the lake to

---

[1] Neither is it of any consequence whether the locality of the
supposed miracle was Gadara, or Gerasa, or Gergesa. But I may
say that I was well acquainted with Origen's opinion respecting
Gergesa. It is fully discussed and rejected in Riehm's *Handwörterbuch*. In Kitto's *Biblical Cyclopædia* (ii. p. 51) Professor
Porter remarks that Origen merely "*conjectures*" that Gergesa
was indicated; and he adds, " Now, in a question of this kind
conjectures cannot be admitted. We must implicitly follow the
most ancient and creditable testimony, which clearly pronounces
in favour of Γαδαρηνῶν. This reading is adopted by Tischendorf,
Alford, and Tregelles."

Gadara had no official status; and they had no more civil right to punish law-breakers than any other strangers.

In my turn, however, I may remark that there is a "point" which appears to have escaped Mr. Gladstone's notice. And that is somewhat unfortunate, because his whole argument turns upon it. Mr. Gladstone assumes, as a matter of course, that pig-keeping was an offence against the "Law of Moses"; and, therefore, that Jews who kept pigs were as much liable to legal pains and penalties as Englishmen who smuggle brandy ("Impregnable Rock," p. 274).

There can be no doubt that, according to the Law, as it is defined in the Pentateuch, the pig was an "unclean" animal, and that pork was a forbidden article of diet. Moreover, since pigs are hardly likely to be kept for the mere love of those unsavoury animals, pig-owning, or swine-herding, must have been, and evidently was regarded as a suspicious and degrading occupation by strict Jews, in the first century A.D. But I should like to know on what provision of the Mosaic Law, as it is laid down in the Pentateuch, Mr. Gladstone bases the assumption, which is essential to his case, that the possession of pigs and the calling of a swineherd were actually illegal. The inquiry was put to me the other day; and, as I could not answer it, I turned up the article "Schwein" in Riehm's standard

"Handwörterbuch," for help out of my difficulty; but unfortunately without success. After speaking of the martyrdom which the Jews, under Antiochus Epiphanes, preferred to eating pork, the writer proceeds :—

> It may be, nevertheless, that the practice of keeping pigs may have found its way into Palestine in the Græco-Roman time, in consequence of the great increase of the non-Jewish popula- tion ; yet there is no evidence of it in the New Testament ; the great herd of swine, 2,000 in number, mentioned in the narrative of the possessed, was feeding in the territory of Gadara, which belonged to the Decapolis ; and the prodigal son became a swineherd with the native of a far country into which he had wandered ; in neither of these cases is there reason for thinking that the possessors of these herds were Jews.[1]

Having failed in my search, so far, I took up the next work of reference at hand, Kitto's "Cyclopædia" (vol. iii. 1876). There, under "Swine," the writer, Colonel Hamilton Smith, seemed at first to give me what I wanted, as he says that swine "appear to have been repeatedly introduced and reared by the Hebrew people,[2] notwithstanding the strong prohibition in the Law of Moses (Is. lxv. 4)." But, in the first place,

---

[1] I may call attention, in passing, to the fact that this author- ity, at any rate, has no sort of doubt of the fact that Jewish Law did not rule in Gadara (indeed, under the head of "Gadara," in the same work, it is expressly stated that the population of the place consisted "predominantly of heathens"), and that he scouts the notion that the Gadarene swineherds were Jews.

[2] The evidence adduced, so far as post-exile times are con- cerned, appears to me insufficient to prove this assertion.

Isaiah's writings form no part of the "Law of Moses"; and, in the second place, the people denounced by the prophet in this passage are neither the possessors of pigs, nor swineherds, but these "which eat swine's flesh and broth of abominable things is in their vessels." And when, in despair, I turned to the provisions of the Law itself, my difficulty was not cleared up. Leviticus xi. 8 (Revised Version) says, in reference to the pig and other unclean animals : "Of their flesh ye shall not eat, and their carcasses ye shall not touch." In the revised version of Deuteronomy, xiv. 8, the words of the prohibition are identical, and a skilful refiner might possibly satisfy himself, even if he satisfied nobody else, that "carcase" means the body of a live animal as well as a dead one ; and that, since swineherds could hardly avoid contact with their charges, their calling was implicitly forbidden. [1] Unfortunately, the authorised version expressly says "dead carcase"; and thus the most rabbinically minded of reconcilers might find his casuistry foiled by that great source of surprises, the "original Hebrew." That such check is at any rate possible, is clear from the fact that the legal uncleanness of some animals, as food, did not interfere with their being lawfully possessed, cared for, and sold by Jews. The

---

[1] Even Leviticus xi. 26, cited without reference to the context, will not serve the purpose ; because the swine *is* "cloven-footed" (Lev. xi. 7).

provisions for the ransoming of unclean beasts
(Lev. xxvii. 27) and for the redemption of their
sucklings (Numbers xviii. 15) sufficiently prove this.
As the late Dr. Kalisch has observed in his ' Commentary " on Leviticus, part ii. p. 129, note :—

> Though asses and horses, camels and dogs, were kept by the
> Israelites, they were, to a certain extent, associated with the
> notion of impurity ; they might be turned to profitable account
> by their labour or otherwise, but in respect to food they were an
> abomination.

The same learned commentator (*loc. cit.* p. 88)
proves that the Talmudists forbade the rearing of
pigs by Jews, unconditionally and everywhere;
and even included it under the same ban as the
study of Greek philosophy, " since both alike were
considered to lead to the desertion of the Jewish
faith." It is very possible, indeed probable, that
the Pharisees of the fourth decade of our first
century took as strong a view of pig-keeping as
did their spiritual descendants. But, for all that,
it does not follow that the practice was illegal.
The stricter Jews could not have despised and
hated swineherds more than they did publicans ;
but, so far as I know, there is no provision in the
Law against the practice of the calling of a tax-
gatherer by a Jew. The publican was in fact
very much in the position of an Irish process-
server at the present day—more, rather than less,
despised and hated on account of the perfect
legality of his occupation. Except for certain

sacrificial purposes, pigs were held in such abhorrence by the ancient Egyptians, that swineherds were not permitted to enter a temple, or to intermarry with other castes; and any one who had touched a pig, even accidentally, was unclean. But these very regulations prove that pig-keeping was not illegal; it merely involved certain civil and religious disabilities. For the Jews, dogs were typically "unclean animals; but, when that eminently pious Hebrew, Tobit, "went forth" with the angel "the young man's dog" went "with them" (Tobit v. 16) without apparent remonstrance from the celestial guide. I really do not see how an appeal to the Law could have justified any one in drowning Tobit's dog, on the ground that his master was keeping and feeding an animal quite as "unclean" as any pig. Certainly the excellent Raguel must have failed to see the harm of dog-keeping, for we are told that, on the travellers' return homewards, "the dog went after them" (xi. 4).

Until better light than I have been able to obtain is thrown upon the subject, therefore, it is obvious that Mr. Gladstone's argumentative house has been built upon an extremely slippery quicksand; perhaps even has no foundation at all.

Yet another "point" does not seem to have occurred to Mr. Gladstone, who is so much shocked that I attach no overwhelming weight to the assertions contained in the synoptic Gospels, even when all three concur. These Gospels agree in stating, in the most express, and to some extent verbally identical, terms, that the devils entered the pigs at their own request,[1] and the third Gospel (viii. 31) tells us what the motive of the demons was in asking the singular boon: "They intreated him that he would not command them to depart into the abyss." From this, it would seem that the devils thought to exchange the heavy punishment of transportation to the abyss for the lighter penalty of imprisonment in swine. And some commentators, more ingenious than respectful to the supposed chief actor in this extraordinary fable, have dwelt, with satisfaction, upon the very unpleasant quarter of an hour which the evil spirits must have had, when the headlong rush of their maddened tenements convinced them how completely they were taken in. In the whole story, there is not one solitary hint that the destruction of the pigs was intended as a punishment of their owners, or of the swineherds. On the contrary, the concurrent testimony of the three narratives is to the effect that the catastrophe was the consequence of diabolic suggestion. And, indeed, no source could

---

[1] 1st Gospel: "And the devils *besought him*, saying, If Thou cast us out send us away *into* the herd of swine." 2d Gospel: "They *besought him*, saying, Send us *into* the swine." 3d Gospel: ' They *intreated him* that he would give them leave to enter *into* them."

be more appropriate for an act of such manifest injustice and illegality.

I can but marvel that modern defenders of the faith should not be glad of any reasonable excuse for getting rid of a story which, if it had been invented by Voltaire, would have justly let loose floods of orthodox indignation.

Thus, the hypothesis, to which Mr. Gladstone so fondly clings, finds no support in the provisions of the " Law of Moses " as that law is defined in the Pentateuch; while it is wholly inconsistent with the concurrent testimony of the synoptic Gospels, to which Mr. Gladstone attaches so much weight. In my judgment, it is directly contrary to everything which profane history tells us about the constitution and the population of the city of Gadara; and it commits those who accept it to a story which, if it were true, would implicate the founder of Christianity in an illegal and inequitable act.

Such being the case, I consider myself excused from following Mr. Gladstone through all the meanderings of his late attempt to extricate himself from the maze of historical and exegetical difficulties in which he is entangled. I content myself with assuring those who, with my paper (not Mr. Gladstone's version of my arguments) in hand, consult the original authorities, that they will find full justification for every statement I have made. But in order to dispose those who cannot, or will not, take that trouble, to believe that the proverbial blindness of one that judges his own cause plays no part in inducing me to speak thus decidedly, I beg their attention to the following examination, which shall be as brief as I can make it, of the seven propositions in which Mr. Gladstone professes to give a faithful summary of my " errors."

When, in the middle of the seventeenth century, the Holy See declared that certain propositions contained in the works of Bishop Jansen were heretical, the Jansenists of Port Royal replied that, while they were ready to defer to the Papal authority about questions of faith and morals, they must be permitted to judge about questions of fact for themselves ; and that, really, the condemned propositions were not to be found in Jansen's writings. As everybody knows, His Holiness and the Grand Monarque replied to this, surely not unreasonable, plea after the manner of Lord Peter in the " Tale of a Tub." It is, therefore, not without some apprehension of meeting with a similar fate, that I put in a like plea against Mr. Gladstone's Bull. The seven propositions declared to be false and condemnable, in that kindly and gentle way which so pleasantly compares with the authoritative style of the Vatican (No. 5 more particularly), may or may not be true. But they are not to be found in anything I have written. And some of them diametrically contravene that which I have written. I proceed to prove my assertions.

PROP. 1. *Throughout the paper he confounds together what I had distinguished, namely, the city of Gadara and the vicinage attached to it, not as a mere pomœrium, but as a rural district.*

In my judgment, this statement is devoid of foundation. In my paper on "The Keepers of the Herd of Swine" I point out, at some length, that, "in accordance with the ancient Hellenic practice," each city of the Decapolis must have been "surrounded by a certain amount of territory amenable to its jurisdiction": and, to enforce this conclusion, I quote what Josephus says about the "villages that belonged to Gadara and Hippos." As I understand the term *pomerium* or *pomœrium*,[1] it means the space which, according to Roman custom, was kept free from buildings, immediately within and without the walls of a city; and which defined the range of the *auspicia urbana*. The conception of a *pomœrium* as a "vicinage attached to" a city, appears to be something quite novel and original. But then, to be sure, I do not know how many senses Mr. Gladstone may attach to the word "vicinage."

Whether Gadara had a *pomœrium*, in the proper technical sense, or not, is a point on which I offer no opinion. But that the city had a very

[1] See Marquardt, *Römische Staatsverwaltung*, Bd. III. p. 408.

considerable "rural district" attached to it and notwithstanding its distinctness, amenable to the jurisdiction of the Gentile municipal authorities, is one of the main points of my case.

PROP. 2. *He more fatally confounds the local civil government and its following, including, perhaps, the whole wealthy class and those attached to it, with the ethnical character of the general population.*

Having survived confusion No. 1, which turns out not to be on my side, I am now confronted in No. 2 with a "more fatal" error—and so it is, if there be degrees of fatality ; but, again, it is Mr. Gladstone's and not mine. It would appear, from this proposition (about the grammatical interpretation of which, however, I admit there are difficulties), that Mr. Gladstone holds that the "local civil government and its following among the wealthy," were ethnically different from the "general population." On p. 348, he further admits that the "wealthy and the local governing power" were friendly to the Romans. Are we then to suppose that it was the persons of Jewish "ethnical character" who favoured the Romans, while those of Gentile "ethnical character" were opposed to them? But, if that supposition is absurd, the only alternative is that the local civil government was ethnically Gentile. This is exactly my contention.

At pp. 379 to 391 of the essay on "The Keepers of the Herd of Swine" I have fully

discussed the question of the ethnical character
of the general population. I have shown that,
according to Josephus, who surely ought to have
known, Gadara was as much a Gentile city as
Ptolemais; I have proved that he includes Gadara
amongst the cities "that rose up against the Jews
that were amongst them," which is a pretty
definite expression of his belief that the "ethnical
character of the general population" was Gentile.
There is no question here of Jews of the Roman
party fighting with Jews of the Zealot party, as
Mr. Gladstone suggests. It is the non-Jewish
and anti-Jewish general population which rises
up against the Jews who had settled "among
them."

PROP. 3. *His one item of direct evidence as to the
Gentile character of the city refers only to the former
and not to the latter.*

More fatal still. But, once more, not to me. I
adduce not one, but a variety of "items" in proof
of the non-Judaic character of the population of
Gadara : the evidence of history; that of the
coinage of the city; the direct testimony of
Josephus, just cited—to mention no others. I
repeat, if the wealthy people and those connected
with them—the "classes" and the "hangers on"
of Mr. Gladstone's well-known taxonomy—were,
as he appears to admit they were, Gentiles; if the
"civil government" of the city was in their hands,
as the coinage proves it was; what becomes of

Mr. Gladstone's original proposition in "The Impregnable Rock of Scripture" that "the population of Gadara, and still less (if less may be) the population of the neighbourhood," were "Hebrews bound by the Mosaic law"? And what is the importance of estimating the precise proportion of Hebrews who may have resided, either in the city of Gadara or in its dependent territory, when, as Mr. Gladstone now seems to admit (I am careful to say "seems"), the government, and consequently the law, which ruled in that territory and defined civil right and wrong was Gentile and not Judaic? But perhaps Mr. Gladstone is prepared to maintain that the Gentile "local civil government" of a city of the Decapolis administered Jewish Law; and showed their respect for it, more particularly, by stamping their coinage with effigies of the Emperors.

In point of fact, in his haste to attribute to me errors which I have not committed, Mr. Gladstone has given away his case.

PROP. 4. *He fatally confounds the question of political party with those of nationality and of religion, and assumes that those who took the side of Rome in the factions that prevailed could not be subject to the Mosaic Law.*

It would seem that I have a feline tenacity of life; once more, a "fatal error." But Mr. Gladstone has forgotten an excellent rule of controversy; say what is true, of course, but mind that it is decently probable. Now it is not decently probable, hardly indeed conceivable, that any one who has read Josephus, or any other historian of the Jewish war, should be unaware that there were Jews (of whom Josephus himself was one) who "Romanised" and, more or less openly, opposed the war party. But, however that may be, I assert that Mr. Gladstone neither has produced, nor can produce, a passage of my writing which affords the slightest foundation for this particular article of his indictment.

PROP. 5. *His examination of the text of Josephus is alike one-sided, inadequate, and erroneous.*

Easy to say, hard to prove. So long as the authorities whom I have cited are on my side, I do not know why this singularly temperate and convincing dictum should trouble me. I have yet to become acquainted with Mr. Gladstone's claims to speak with an authority equal to that of scholars of the rank of Schürer, whose obviously just and necessary emendations he so unceremoniously pooh-poohs.

PROP. 6. *Finally, he sets aside, on grounds not critical or historical, but partly subjective, the primary historical testimony on the subject, namely, that of the three Synoptic Evangelists, who write as contemporaries and deal directly with the subject, neither of which is done by any other authority.*

Really this is too much !   The fact is, as anybody

can see who will turn to my article of February 1889 [VII. *supra*], out of which all this discussion has arisen, that the arguments upon which I rest the strength of my case touching the swine-miracle, are exactly " historical " and " critical." Expressly, and in words that cannot be misunderstood, I refuse to rest on what Mr. Gladstone calls " subjective " evidence. I abstain from denying the possibility of the Gadarene occurrence, and I even go so far as to speak of some physical analogies to possession. In fact, my quondam opponent, Dr. Wace, shrewdly, but quite fairly, made the most of these admissions; and stated that I had removed the only " consideration which would have been a serious obstacle " in the way of his belief in the Gadarene story. [1]

So far from setting aside the authority of the synoptics on " subjective " grounds, I have taken a great deal of trouble to show that my non-belief in the story is based upon what appears to me to be evident; firstly, that the accounts of the three synoptic Gospels are not independent, but are founded upon a common source; secondly, that, even if the story of the common tradition proceeded from a contemporary, it would still be worthy of very little credit, seeing the manner in which the legends about mediæval miracles have been propounded by contemporaries. And in

[1] *Nineteenth Century*, March 1889 (p. 362).

illustration of this position I wrote a special essay about the miracles reported by Eginhard. [1]

In truth, one need go no further than Mr. Gladstone's sixth proposition to be convinced that contemporary testimony, even of well-known and distinguished persons, may be but a very frail reed for the support of the historian, when theological prepossession blinds the witness.[2]

PROP. 7. *And he treats the entire question, in the narrowed form in which it arises upon secular testimony, as if it were capable of a solution so clear and*

[1] "The Value of Witness to the Miraculous." *Nineteenth Century*, March 1889.

[2] I cannot ask the Editor of this Review to reprint pages of an old article,—but the following passages sufficiently illustrate the extent and the character of the discrepancy between the facts of the case and Mr. Gladstone's account of them :—

"Now, in the Gadarene affair, I do not think I am unreasonably sceptical if I say that the existence of demons who can be transferred from a man to a pig does thus contravene probability. Let me be perfectly candid. I admit I have no *à priori* objection to offer. . . . I declare, as plainly as I can, that I am unable to show cause why these transferable devils should not exist." . . . ("Agnosticism," *Nineteenth Century*, 1889, p. 177).

"What then do we know about the originator, or originators, of this groundwork—of that threefold tradition which all three witnesses (in Paley's phrase) agree upon—that we should allow their mere statements to outweigh the counter arguments of humanity, of common sense, of exact science, and to imperil the respect which all would be glad to be able to render to their Master?" (*ibid.* p. 175).

I then go on through a couple of pages to discuss the value of the evidence of the synoptics on critical and historical grounds. Mr. Gladstone cites the essay from which these passages are taken, whence I suppose he has read it ; though it may be that he shares the impatience of Cardinal Manning where my writings are concerned. Such impatience will account for, though it will not excuse, his sixth proposition.

summary as to warrant the use of the extremest weapons of controversy against those who presume to differ from him.

The six heretical propositions which have gone before are enunciated with sufficient clearness to enable me to prove, without any difficulty, that, whosesoever they are, they are not mine. But number seven, I confess, is too hard for me. I cannot undertake to contradict that which I do not understand.

What is the "entire question" which "arises" in a "narrowed form" upon "secular testimony"? After much guessing, I am fain to give up the conundrum. The "question" may be the ownership of the pigs; or the ethnological character of the Gadarenes; or the propriety of meddling with other people's property without legal warrant. And each of these questions might be so "narrowed" when it arose on "secular testimony" that I should not know where I was. So I am silent on this part of the proposition.

But I do dimly discern, in the latter moiety of this mysterious paragraph, a reproof of that use of "the extremest weapons of controversy" which is attributed to me. Upon which I have to observe that I guide myself, in such matters, very much by the maxim of a great statesman, "Do ut des." If Mr. Gladstone objects to the employment of such weapons in defence, he would do well to abstain from them in attack. He should not frame charges which he has, afterwards, to admit are erroneous, in language of carefully calculated offensiveness ("Impregnable Rock," pp. 269-70); he should not assume that persons with whom he disagrees are so recklessly unconscientious as to evade the trouble of inquiring what has been said or known about a grave question ("Impregnable Rock," p. 273); he should not qualify the results of careful thought as "hand-over-head reasoning" ("Impregnable Rock," p. 274); he should not, as in the extraordinary propositions which I have just analysed, make assertions respecting his opponent's position and arguments which are contradicted by the plainest facts.

Persons who, like myself, have spent their lives outside the political world, yet take a mild and philosophical concern in what goes on in it, often find it difficult to understand what our neighbours call the psychological moment of this or that party leader, and are, occasionally, loth to believe in the seeming conditions of certain kinds of success. And when some chieftain, famous in political warfare, adventures into the region of letters or of science, in full confidence that the methods which have brought fame and honour in his own province will answer there, he is apt to forget that he will be judged by these people, on whom rhetorical artifices have long ceased to take effect; and to whom mere dexterity in putting together cleverly ambiguous phrases, and even the great art of offensive misrepresentation, are unspeakably wearisome. And, if that weariness finds its expression in sarcasm, the offender really has no right to cry out. Assuredly, ridicule is no test of truth, but it is the righteous meed of some kinds of error. Nor ought the attempt to confound the expression of a revolted sense of fair dealing with arrogant impatience of contradiction, to restrain those to whom " the extreme weapons of controversy " come handy from using them. The function of police in the intellectual, if not in the civil, economy may sometimes be legitimately discharged by volunteers.

Some time ago, in one of the many criticisms with which I am favoured, I met with the remark that, at our time of life, Mr. Gladstone and I might be better occupied than in fighting over the Gadarene pigs. And, if these too famous swine were the only parties to the suit, I, for my part, should fully admit the justice of the rebuke. But, under the beneficent rule of the Court of Chancery, in former times, it was not uncommon, that a quarrel about a few perches of worthless land, ended in the ruin of ancient families and the engulfing of great estates; and I think that our admonisher failed to observe the analogy—to note the momentous consequences of the judgment

which may be awarded in the present apparently insignificant action *in re* the swineherds of Gadara.

The immediate effect of such judgment will be
the decision of the question, whether the men of
the nineteenth century are to adopt the demonology of the men of the first century, as divinely
revealed truth, or to reject it, as degrading falsity.
The reverend Principal of King's College has
delivered his judgment in perfectly clear and
candid terms. Two years since, Dr. Wace said
that he believed the story as it stands; and consequently he holds, as a part of divine revelation,
that the spiritual world comprises devils, who,
under certain circumstances, may enter men and
be transferred from them to four-footed beasts.
For the distinguished Anglican Divine and Biblical
scholar, that is part and parcel of the teachings
respecting the spiritual world which we owe to the
founder of Christianity. It is an inseparable part
of that Christian orthodoxy which, if a man
rejects, he is to be considered and called an
"infidel." According to the ordinary rules of
interpretation of language, Mr. Gladstone must
hold the same view.

If antiquity and universality are valid tests of
the truth of any belief, no doubt this is one of the
beliefs so certified. There are no known savages,
nor people sunk in the ignorance of partial civilisation, who do not hold them. The great majority

of Christians have held them and still hold them. Moreover the oldest records we possess of the early conceptions of mankind in Egypt and in Mesopotamia prove that exactly such demonology, as is implied in the Gadarene story, formed the substratum, and, among the early Accadians, apparently the greater part. of their supposed knowledge of the spiritual world. M. Lenormant's profoundly interesting work on Babylonian magic and the magical texts given in the Appendix to Professor Sayce's "Hibbert Lectures" leave no doubt on this head. They prove that the doctrine of possession, and even the particular case of pig possession,[1] were firmly believed in by the Egyptians and the Mesopotamians before the tribes of Israel invaded Palestine. And it is evident that these beliefs, from some time after the exile and probably much earlier, completely interpenetrated the Jewish mind, and thus became inseparably interwoven with the fabric of the synoptic Gospels.

Therefore, behind the question of the acceptance of the doctrines of the oldest heathen demonology as part of the fundamental beliefs of Christianity, there lies the question of the credibility of the

---

[1] The wicked, before being annihilated, returned to the world to disturb men ; they entered into the body of unclean animals, "often that of a pig, as on the Sarcophagus of Seti I. in the Soane Museum."—Lenormant, *Chaldean Magic*, p. 88, Editorial Note.

Gospels, and' of their claim to act as our instruct-
ors, outside that ethical province in which they
appeal to the consciousness of all thoughtful men.
And still, behind this problem, there lies another
—how far do these ancient records give a sure
foundation to the prodigious fabric of Christian
dogma, which has been built upon them by the
continuous labours of speculative theologians,
during eighteen centuries ?

I submit that there are few questions before
the men of the rising generation, on the answer
to which the future hangs more fatally, than this.
We are at the parting of the ways. Whether the
twentieth century shall see a recrudescence of the
superstitions of mediæval papistry, or whether it
shall witness the severance of the living body of
the ethical ideal of prophetic Israel from the carcase, foul with savage superstitions and cankered
with false philosophy, to which the theologians
have bound it, turns upon their final judgment of
the Gadarene tale.

The gravity of the problems ultimately involved
in the discussion of the legend of Gadara will, I
hope, excuse a persistence in returning to the subject, to which I should not have been moved by
merely personal considerations.

With respect to the diluvial invective which
overflowed thirty-three pages of the " Nineteenth

Century "last January, I doubt not that it has a catastrophic importance in the estimation of its author. I, on the other hand, may be permitted to regard it as a mere spate; noisy and threatening while it lasted, but forgotten almost as soon as it was over. Without my help, it will be judged by every instructed and clear-headed reader; and that is fortunate, because, were aid necessary, I have cogent reasons for withholding it.

In an article characterised by the same qualities of thought and diction, entitled "A Great Lesson," which appeared in the "Nineteenth Century" for September 1887, the Duke of Argyll, firstly, charged the whole body of men of science, interested in the question, with having conspired to ignore certain criticisms of Mr. Darwin's theory of the origin of coral reefs; and, secondly, he asserted that some person unnamed had "actually induced" Mr. John Murray to delay the publication of his views on that subject "for two years."

It was easy for me and for others to prove that the first statement was not only, to use the Duke of Argyll's favourite expression, "contrary to fact," but that it was without any foundation whatever. The second statement rested on the Duke of Argyll's personal authority. All I could do was to demand the production of the evidence for it. Up to the present time, so far as I know, that evidence has not made its appearance; nor has there been any withdrawal of, or apology for, the erroneous charge.

Under these circumstances most people will understand why the Duke of Argyll may feel quite secure of having the battle all to himself, whenever it pleases him to attack me.

[See the note at the end of "Hasisadra's Adventure" (vol iv. p. 283). The discussion on coral reefs, at the meeting of the British Association this year, proves that Mr. Darwin's views are defended now, as strongly as in 1891, by highly competent authorities.   October 25, 1893.]

END OF VOL. V

RICHARD CLAY AND SONS, LIMITED,
LONDON AND BUNGAY.

Milton Keynes UK
Ingram Content Group UK Ltd.
UKHW032320161024
449665UK00001B/35